CAMBRIDGE LIBRARY COLLECTION

Books of enduring scholarly value

Life Sciences

Until the nineteenth century, the various subjects now known as the life
sciences were regarded either as arcane studies which had little impact
on ordinary daily life, or as a genteel hobby for the leisured classes. The
increasing academic rigour and systematisation brought to the study of
botany, zoology and other disciplines, and their adoption in university
curricula, are reflected in the books reissued in this series.

The Wild Garden

An Irish-born gardener and writer, William Robinson (1838–1935) travelled
widely to study gardens and gardening in Europe and America. He founded
a weekly illustrated periodical, *The Garden*, in 1871, which he owned until
1919, and published numerous books on different aspects of horticulture. His
most famous book, *The English Flower Garden* (also reprinted in this series),
was published in 1883, and fifteen editions were issued in his lifetime. *The
Wild Garden*, published in 1870, attacks contemporary fashions in public
parks and private gardens, which involved showy masses of colour in labour-
intensive summer bedding, using mostly subtropical and exotic species.
He calls for a return to native species, found in traditional English gardens,
pointing out that these are more economical than short-lived annuals,
and that there is much greater variety available to the garden designer. He
suggests plants more suitable for the English climate, and exotics more
capable of naturalisation.

Cambridge University Press has long been a pioneer in the reissuing of out-of-print titles from its own backlist, producing digital reprints of books that are still sought after by scholars and students but could not be reprinted economically using traditional technology. The Cambridge Library Collection extends this activity to a wider range of books which are still of importance to researchers and professionals, either for the source material they contain, or as landmarks in the history of their academic discipline.

Drawing from the world-renowned collections in the Cambridge University Library, and guided by the advice of experts in each subject area, Cambridge University Press is using state-of-the-art scanning machines in its own Printing House to capture the content of each book selected for inclusion. The files are processed to give a consistently clear, crisp image, and the books finished to the high quality standard for which the Press is recognised around the world. The latest print-on-demand technology ensures that the books will remain available indefinitely, and that orders for single or multiple copies can quickly be supplied.

The Cambridge Library Collection will bring back to life books of enduring scholarly value (including out-of-copyright works originally issued by other publishers) across a wide range of disciplines in the humanities and social sciences and in science and technology.

The Wild Garden

*Or, Our Groves and Shrubberies
Made Beautiful*

WILLIAM ROBINSON

CAMBRIDGE UNIVERSITY PRESS

Cambridge, New York, Melbourne, Madrid, Cape Town,
Singapore, São Paolo, Delhi, Tokyo, Mexico City

Published in the United States of America by Cambridge University Press, New York

www.cambridge.org
Information on this title: www.cambridge.org/9781108037105

This edition first published 1870
This digitally printed version 2011

ISBN 978-1-108-03710-5 Paperback

THE WILD GARDEN.

"I wish it to be framed, as much as may be, to a naturall wildnesse."

LORD BACON.

THE WILD GARDEN

OR,

OUR GROVES & SHRUBBERIES

MADE BEAUTIFUL

BY THE NATURALIZATION OF HARDY EXOTIC PLANTS:

WITH A CHAPTER

ON THE GARDEN OF BRITISH WILD FLOWERS.

By W. ROBINSON,

AUTHOR OF

"ALPINE FLOWERS FOR ENGLISH GARDENS," "THE PARKS, PROMENADES,
AND GARDENS OF PARIS," ETC.

LONDON:
JOHN MURRAY, ALBEMARLE STREET.
1870.

LONDON :
SAVILL, EDWARDS AND CO., PRINTERS, CHANDOS STREET,
COVENT GARDEN.

CONTENTS.

―――

PART I.

PART II.

PART III.

PART IV.

"I WENT, for the first time in my life, some years ago, to stay at a very
grand and beautiful place in the country, where the grounds are said to
be laid out with consummate taste. For the first three or four days I
was perfectly enchanted; it seemed something so much better than
nature that I really began to wish the earth had been laid out according
to the latest principles of improvement. . . . In three days' time I
was tired to death: a thistle, a nettle, a heap of dead bushes—anything
that wore the appearance of accident and want of intention—was quite
a relief. I used to escape from the made grounds, and walk upon an
adjacent goose-common, where the cart-ruts, gravel-pits, bumps, irregu-
larities, coarse ungentlemanlike grass, and all the varieties produced by
neglect, were a thousand times more gratifying than the monotony of
beauties the result of design, and crowded into narrow confines."

SYDNEY SMITH.

PART I.

EXPLANATORY.

THE WILD GARDEN.

To understand the aim of this little book, it is desirable to take a broad glance at the past and present state of our flower-gardens. From about twenty years ago, back to the time of Shakespeare, the flowers of an English garden were nearly all hardy ones : they came from northern or temperate regions, in most cases from climates very like our own ; they were as hardy as our weeds ; they bloomed early in the keen spring air, and late in the wet autumn gusts, as well as in the favoured summer's day.

> The daughters of the year,
> One after one, thro' that still garden passed.

Passages from our greatest poets and writers— Shakespeare, Milton, Bacon, and others—embody the names of the principal classes of flowers used in this ancient style of gardening, and show us what infinite delight it was capable of affording ; and its

charms we may yet see in little cottage gardens in
Kent, Sussex, and many other parts of England,
though the scarlet geranium has begun to eradicate
all the fair blossoms of many a sweet little garden,
once, and often yet, " embowered in fruit trees and
forest trees, evergreens and honeysuckles rising
many-coloured from amid shaven grass plots, flowers
struggling in through the very windows . . . where,
especially on long summer nights, a king might
have wished to sit and smoke and call it his." From
these little Elysiums, where the last glimpses of
beautiful old English gardening may yet be seen,
we will now turn to the modern system which re-
places it.

About a generation ago a taste began to be
manifested for placing a number of tender plants in
the open air in summer, with a view to the production of showy masses of decided colour.　The subjects selected were mostly from sub-tropical climates
and of free growth ; placed in the open air of our
genial early summer, and in fresh rich earth, every
year they grew rapidly and flowered abundantly
during the summer and early autumn months,
and until cut down by the first frosts.　The bril-
liancy of tone resulting from this system was very
attractive, and since its introduction there has

been a gradual rooting out of all the old favourites
in favour of the bedding system. This was carried
to such an extent that of late it has not been uncommon, indeed it has been the rule, to find the
largest gardens in the country without a single hardy
flower, all energy and expense being devoted to
the production of the many thousand exotics required for the summer decoration. It should be
distinctly borne in mind that the expense for this
system is an annual one ; that no matter what
amount of money may be spent in this way, no
matter how many years may be devoted to perfecting it, the first sharp frost of November merely
prepares a yet further expense and labour.

Its highest results need hardly be described ;
they are seen in all our great public gardens ;
our London and many other city parks show them
in the shape of beds filled with vast quantities
of flowers, covering the ground frequently in a
showy way, and not unfrequently in a repul-
sively gaudy manner : every private garden is
taken possession of by the same simple beauties.
Occasionally some variety is introduced. We go
to Kew or the Crystal Palace to see what looks
best there, or the weekly gardening papers tell us ;
and the following season sees tens of thousands of

the same arrangements and patterns scattered all over the country. I will not here enter into the question of the comparative advantages of the two systems; it is enough to state that even on its votaries the system at present in fashion is beginning to pall. Some are looking back with regret to the old mixed-border gardens ; others are endeavouring to soften the harshness of the bedding system by the introduction of fine-leaved plants, but all are agreed that a great mistake has been made in destroying all our sweet old border flowers, from tall Lilies to dwarf Hepaticas, though very few persons indeed have any idea of the numbers of beautiful subjects in this way which we may gather from every northern and temperate clime.

What is to be done ? Every garden should have a mixed border, but except in the little cottage gardens before alluded to—"umbrageous man's nests," as Mr. Carlyle calls them, gardens dependent on it solely are quite out of the question. It is also clear that, base and frightfully opposed to every law of nature's own arrangement of living things as is the bedding system, it has yet some features which deserve to be retained on a small scale. My object is now to show how we may, without losing the better features of the mixed

bedding or any other system, follow one infinitely
superior to any now practised, yet supplementing
both, and exhibiting more of the varied beauty of
hardy flowers than the most ardent admirer of the
old style of garden ever dreams of. We may do
this by naturalizing or making wild innumerable
beautiful natives of many regions of the earth in
our woods, wild and semi-wild places, rougher parts
of pleasure grounds, etc., and in unoccupied places
in almost every kind of garden.

I allude not to the wood and brake flora of any
one alp or chain of alps, but to that which finds its
home in the immeasurable woodlands that fall in
furrowed folds from beneath the hoary heads of all
the great mountain chains of the world, whether
they rise from hot Indian plains or green European
pastures. The Palm and sacred Fig, as well as the
Wheat and the Vine, are separated from the stemless plants that cushion under the snow for half the
year, by a zone of hardier and not less beautiful life,
varied as the breezes that whisper on the mountain
sides, and as the little rills that seam them. I allude to
the Lilies, and Bluebells, and Foxgloves, and Irises,
and Windflowers, and Columbines, and Aconites, and
Rock-roses, and Violets, and Cranesbills, and count-
less Pea-flowers, and mountain Avens, and Brambles,

and Cinquefoils, and Evening Primroses, and Clematises, and Honeysuckles, and Michaelmas Daisies,
and Feverfews, and Wood-hyacinths, and Daffodils,
and Bindweeds, and Forget-me-nots, and sweet
blue Omphalodes, and Primroses, and Day Lilies,
and Asphodels, and St. Bruno's Lilies, and the
almost innumerable plants which form the flora of
regions where, though life is yet rife on every inch
of ground, and we are enjoying the verdure and the
temperature of our lowland meadows, there is
a "sense of a great power beginning to be manifested in the earth, and of a deep and majestic
concord in the rise of the long low lines of piny
hills; the first utterances of those mighty mountain symphonies, soon to be more loudly lifted
and wildly broken along the battlements of the
Alps. But their strength is as yet restrained, and
the far-reaching ridges of pastoral mountains succeed
each other, like the long and sighing swell which
moves over quiet waters, from some far-off stormy
sea. And there is a deep tenderness pervading
that vast monotony. The destructive forces, and
the stern expression of the central ranges, are alike
withdrawn. No frost-ploughed, dust-encumbered
paths of the ancient glacier fret the soft Jura pastures; no splintered heaps of ruin break the fair

ranks of her forests ; no pale, defiled, or furious rivers rend their rude and changeful ways among her rocks. Patiently, eddy by eddy, the clear green streams wind along their well - known beds; and under the dark quietness of the undisturbed pines there spring up, year by year, such company of joyful flowers as I know not the like of among all the blessings of the earth. It was spring-time, too ; and all were coming forth in clusters crowded for very love. There was room enough for all, but they crushed their leaves into all manner of strange shapes, only to be nearer each other. There was the Wood Anemone, star after star, closing every now and then into nebulæ ; and there was the Oxalis, troop by troop, like virginal processions of the Mois de Marie, the dark vertical clefts in the limestone choked up with them as with heavy snow, and touched with Ivy on the edges— Ivy as light and lovely as the Vine ; and, ever and anon, a blue gush of Violets and Cowslip bells in sunny places ; and in the more open ground, the Vetch, and Comfrey, and Mezereon, and the small sapphire buds of the alpine Polygala, and the Wild Strawberry, just a blossom or two, all showered amidst the golden softness of deep, warm, amber-coloured moss."

This is a picture of but one of innumerable and infinitely varied scenes in the wilder parts of all northern and temperate regions, at many different elevations. The loveliness and ceaselessly varying charms of such scenes are indeed difficult to describe or imagine ; the essential thing to bear in mind is that the plants that go to form them *are hardy, and will thrive in our climate as well as native plants.*

Such beauty may be realized in every wood and copse and wild shrubbery that screens our "trim gardens." Naturally our woods and wilds have no small loveliness in spring ; we have here and there the Lily-of-the-valley and the Snowdrop wild, and everywhere the exquisite Primrose and Cowslip ; the Bluebell and the Foxglove sometimes take nearly complete possession of whole woods, and turn them into paradises of vernal beauty ; but, with all our treasures in this way, we have no attractions in semi-wild places compared to what it is within our power to create. A certain number of beautiful plants occur amongst the weeds in our woods, and there we stop. But there are many countries with winters as cold as, or colder than, our own, possessing a rich flora ; and by taking the best hardy exotics and establishing them with the best of our own wild

flowers in wild or half-wild spots near our houses
and gardens, we may produce the most charming
results ever seen in such places. To most people a
pretty plant in the wild state is more attractive than
any garden denizen. It is free, and taking care of
itself, it has had to contend with and has over-
come weeds which, left to their own sweet will
in a garden, would soon leave very small trace
of the plants therein ; and, moreover, it is usually
surrounded by some degree of graceful wild spray
—the green above, and the moss and brambles
and grass around. Many will say with Tennyson,
in " Amphion,"—

> Better to me the meanest weed
> That blows upon its mountain,
> The vilest herb that runs to seed
> Beside its native fountain—

but by the means presently to be explained, numbers of plants, neither " mean " nor " vile," but of
the highest order of beauty and fragrance, and
clothed with the sweetest associations, may be seen
to greater perfection, wild as weeds, in the spaces
now devoted to rank grass and weeds in our shrubberies, ornamental plantations, and by wood walks,
than ever they were in our gardens.

My reasons for advocating this system, as I do,

are as follows : *first,* because hundreds of the finest
hardy flowers will thrive much better in the places
I recommend for them than ever they did in the
old-fashioned border.   Even comparatively small
ones, like the ivy-leaved Cyclamen, a beautiful
plant that we rarely find in perfection in gardens,
I have seen perfectly naturalized and spread all
over the mossy surface of a thin wood.   *Secondly,*
because they will look infinitely better than ever
they did in gardens, in consequence of fine-leaved
plant, fern, and flower, and climber, ornamental
grass and dwarf trailing shrub, mutually relieving
each other in ways innumerable as delightful.   Any
one of a thousand combinations, which this book
will suggest to the intelligent reader, will prove as
far superior to any aspect of the old mixed border,
or the ordinary type of modern flower-garden, as is
a lovely mountain valley to a country in which the
eye can see but canals and hedges.   *Thirdly,* because, arranged as I propose, no disagreeable effects
result from decay.   The raggedness of the old mixed
border after the first flush of spring and early summer bloom had passed was intolerable, bundles of
decayed stems tied to sticks making the place look
like the parade-ground of a number of crossing-
sweepers with their " arms piled."   When Lilies are

sparsely dotted through masses of Rhododendrons as I recommend, their flowers are admired more than if they were in isolated showy masses; when they pass out of bloom they are unnoticed amidst the vegetation, and not eyesores, as when in rigid unrelieved tufts in borders, &c. In a wild or semi-wild state, the beauty of individual species will proclaim itself when at its height; and when passed out of bloom, they will be succeeded by other kinds, or lost among the numerous objects around *Fourthly*, because it will enable us to grow hundreds of plants that have never yet obtained a place in our " trim gardens," nor ever will be admitted therein. I allude to the multitudes of plants which, not being so showy as those usually considered worthy of a place in gardens, are never seen there. The flowers of many of these are of the highest order of beauty, especially when seen in numbers. An isolated tuft of one of these, seen in a formal border, may not be considered worthy of a place at any time—in some wild glade, in a wood, associated with other subjects, its effect may be exquisite. We do not usually cultivate Gorse or Buttercups, yet Mr. Wallace, the distinguished naturalist and traveller, says—" During twelve years spent amidst the grandest tropical vegetation, I have

seen nothing comparable to the effect produced on
our landscapes by Gorse, Broom, Heather, Wild
Hyacinths, Hawthorn, and Buttercups;" and
these are but a few conspicuous members of our
indigenous flora, which is by no means as rich as
those of many other cold countries! In every
county in the British Isles there are numbers of
country seats in which one hundred types of vegetation, novel, yet as beautiful as, or more beautiful
than, those admired by Mr. Wallace, may be established ; for there are in the colder parts of Europe,
Asia, and other countries, Heaths handsomer than
those usually grown, many "wild Hyacinths" besides the common English one, many finer "Buttercups" than those commonly seen, and numbers of
Hawthorns besides our common May ; not to speak
of many other families and plants equally beautiful.
Among the subjects that are usually considered
unfit for garden cultivation may be included a
goodly number that, grown in gardens, are little
addition to them ; I mean subjects like the American
Asters, Golden Rods, and like plants, which merely
tend to hide the beauty of the choicer and more
beautiful border-flowers when planted amongst them.
These coarse subjects would be quite at home in
copses and woody places, where their blossoms

might be seen or gathered in due season, and their vigorous vegetation form a covert welcome to the game preserver. To these two groups might be added subjects like the winter Heliotropes, the handsome British Epilobium angustifolium, and many other plants which, while attractive in the garden, are apt to spread about so rapidly as to become a nuisance there. Clearly these should only be planted in wild and semi-wild places. *Fifthly*, because we may in this way settle also the question of spring flowers, and the spring garden, as well as that of hardy flowers generally. In the way I suggest, many parts of every country garden, and many suburban ones, may be made alive with spring flowers. The blue stars of the Apennine Anemone will be seen to greater advantage " wild," in shady or half-shady bare places, under trees, than in any conceivable formal arrangement, and it is but one of hundreds of sweet spring flowers that will succeed perfectly in the way I propose. *Sixthly*, because there can be few more agreeable phases of communion with nature than naturalizing the natives of countries in which we are infinitely more interested than in those of greenhouse or stove plants. From the walls of the Coliseum, the prairies of the New World, the woods and meadows of all

the great mountains of Europe; from Greece and
Italy and Spain, from the sunny hills of Asia
Minor; from the arctic regions of the great conti-
nents—in a word, from almost every region interesting to the traveller he may bring seeds or plants
and establish round his home the pleasantest
souvenirs of the various scenes he has visited.

Moreover, the great merit of permanence belongs
to this delightful phase of gardening.   Select a wild
rough slope, and embellish it with the handsomest
and hardiest climbing plants,—say the noble mountain Clematis from Nepal, the sweet C. Flammula
from Southern Europe, and the magnificent new
hybrid Clematises, (if the earth be rich and there
are rocks and banks on which they can be so
arranged that they will not be overrun by coarser
kinds, and that their masses of shoots may spread
and bask in the sun till they glow into sheets of
purple of various shades) "Virginian creepers" in
variety, Rubus biflorus, with its whitewashed stems,
and other kinds; various species of hardy vines,
Aristolochias, Jasmines, Honeysuckles — British
and European, wild Roses, etc.   Arranged with
some judgment at first, such a colony might be left
to take care of itself; time would but add to its
attractions, and the owner might go away for ten

years, and find it more beautiful than ever on his
return. As much may be said of all the other combinations which I suggest.

I will now endeavour to illustrate my meaning
by showing what may be done with a few diverse
types of northern vegetation. We will take the
Forget-me-not order to begin with, and as that is
one far from being as rich as others in subjects
suited for naturalization, the reader may be able to
form some idea of what we may do, in this way, by
selecting from the numerous families of plants that
grow in the meadows and mountain-woods of
Europe, Asia, and America.

The Forget-me-not or Borage family is a well-
marked and well-known one, containing a great
number of coarse and ugly weeds, but which, if it included only the common Forget-me-not among its
beauties, would have some claims to our attention.
Many persons are not acquainted with more than a
couple of the Forget-me-nots; but what lovely
exotic plants there are in this order that would
afford delight if met with creeping about along
our wood and shrubbery walks! Nature, say some,
is sparing of her deep true blues, and generally
spreads them forth on the high Alps, where the
Gentians bloom near to the sky; but there are

C

obscure plants in this order that possess the truest,
deepest, and most delicate of blues, and which will
thrive as well in the positions I allude to as common
weeds. The Gentians and high alpine plants require
some care in our sluggish lowlands, but not so these.
The creeping Omphalodes verna even surpasses the
Forget-me-not in the depth and beauty of its blue
and its general good qualities, and runs about quite
freely in any shady or half-shady shrubbery, wood,
or rough rockwork. Its proper home is the wood
or semi-wild spot, where it takes care of itself. Put
it in a garden, and probably, unless the soil and region be moist, it soon perishes. Besides, in the
border, it would be a not very agreeable object
when once the sweet spring bloom had passed ;
whereas in the positions spoken of, in consequence
of the predominance of trees, shrubs, and tall herbs,
the low plants are not noticed when out of flower,
but crawl about unobserved till returning spring re-
minds those fortunate enough to see them how
chaste and superior is the inexpensive and natural
kind of gardening here advocated.

Another plant of the order is so suitable and useful for this purpose, that if a root or two of it be
planted in any shrubbery, it will soon run about,
exterminate the weeds, and prove quite a lesson

in wild and natural gardening. I allude to the beautiful Caucasian Comfrey (Symphytum caucasicum), which grows about twenty inches high, and bears quantities of the loveliest blue pendulous flowers. It, like many others, does much better in a wood, grove, or any kind of shrubbery, than in any other position, just filling in the naked spaces between the trees and shrubs, and has a quick-growing and spreading tendency, but never becomes weedy or objectionable. As if to contrast with it, there is the deep crimson Bohemian Comfrey (S. bohemicum), which is sometimes startling from the depth of its vivid colouring, and the white Comfrey (S. orientale), quite a vigorous-growing kind, blooming early in April and May, with the blue Caucasian C.

I purposely omit the British Forget-me-nots, wishing now chiefly to show what we may do with exotics quite as hardy as our own wildings ; and we have another Forget-me-not, not British, which surpasses them all — the early Myosotis dissitiflora. This is like a patch of the bluest sky settled down among the moist stones of a rockwork or any similar spot before our own Forget-me-not has opened its blue eyes, and is admirable for glades or banks in wood or shrubbery, especially in moist districts.

For rocky bare places and sunny sandy banks
we have the spreading Gromwell (Lithospermum
prostratum), which, when in flower, looks just as if
some exquisite alpine Gentian had assumed the
form of a matted hispid bush, to enable it to hold
its own among creeping things and stouter herbs
than accompany it on the Alps. Also the dwarf
spring-blooming Lungworts (Pulmonarias), the
handsome profuse-flowering Italian Bugloss (Anchusa), and the Apennine Hounds-tongue (Cynoglossum), and that strong old plant the Cretan Borage
(generally known as Nordmannia cordifolia), which
opens its lavender-blue and conspicuous flowers in
early spring, and is tall and strong enough to maintain its position even among Docks or Nettles. It
would be found to delight in any old lane or by-
path with the winter Heliotrope or the like, while
there would be no fear of its becoming a weed, like
that sweet-scented wilding.

We will next turn from the Forget-me-not order
to a very different type of vegetation—hardy bulbs.
How many of us really enjoy the beauty which a judicious use of a profusion of good and cheap Spring
Bulbs is certain to throw around a country seat or villa
garden ? How many get beyond the miserable con-
ventionalities of modern gardening, with its edgings

and patchings, and taking up, and drying, and mere playing with our beautiful Spring Bulbs? How many enjoy the exquisite beauty afforded by Spring flowers of this type, established naturally, and cropping up full of beauty, without troubling us for attention at any time? None. The subject of decorating with Spring Bulbs is merely in its infancy; at present we merely place a few of the showiest of them in geometrical lines. The little we do leads to such a very poor end, that numbers of people, alive to the real charms of a garden too, scarcely notice Spring Bulbs at all, regarding them as things which require endless trouble, as interfering with the "bedding-out," and in fact, as not worth the pains they occasion. This is likely to be the case so long as the most effective and satisfactory of all modes of arranging them is quite unused by the body of the gardening public; that way is the placing of them in wild and semi-wild parts of country seats and gardens, and in the rougher parts of a garden, no matter where it may be situated or how it may be arranged. It is a way never practised now, but which I venture to say will yield more real interest and exquisite beauty than any other.

Look, for instance, at the wide and bare belts of grass that wind in and around the shrubberies in nearly every country place ; generally, they never display a particle of plant-beauty, and are merely places to be roughly mown now and then. But if planted here and there with the Snowdrop, the blue Anemone, the Crocus, Squills, and Winter Aconite, they would in spring surpass in attractiveness to the tasteful eye the primmest and gayest of spring gardens. Cushioned among the grass, these would have a more congenial medium in which to unfold than is offered by the beaten sticky earth of a border : in the budding emerald grass of spring, their natural bed, they would look far better than ever they do when arranged on the brown earth of a garden. Once carefully planted, they—while an annual source of the greatest interest—occasion no trouble whatever. Their leaves die down so early in spring that they would scarcely interfere with the mowing of the grass, if that were desired, but I should not attempt to mow the grass in such places till the season of vernal beauty had quite passed by.

Surely it is enough to have the lawn as smooth as a carpet at all times, without sending the mower to shave the "long and pleasant grass" of the remoter parts of the grounds. It would indeed be well worth while to leave many parts of the grass

unmown for the sake of growing Spring Bulbs.
Observe how the poet's eye is caught by the
buttercups that "shine like gold" there ; and we,
who are continually talking of our "horticultural
skill and progress," never so much as get near the
effect produced by this very glinting field of buttercups, or attain to anything which at all equals it in
beauty, although our opportunities to do so are un-
rivalled ! Now suppose a poet, with an eye for natural beauty, or an artist, or any person of taste, to
come upon some spot where a wide fringe of grass
spreads out in the bay of a shrubbery or plantation,
and upon this carpet of rising and unshaven verdure
there were dotted, in addition to the few pretty
natural flowers that happened to take possession of
it, the blue Apennine Anemone, the Snowdrop,
Crocuses, "both the yellow and the gray," as Lord
Bacon has it, Scillas in variety, Grape Hyacinths,
Wood Anemone, and any other pretty Spring
flowers that you found suitable to your soil and
position—say, for instance, a sprinkling of the
Sweet Violet — what would you have done for
him here ? Why, more than the gardener has
ever yet accomplished, because you would have
given him a glimpse of the choicest vernal beauty
of temperate and northern climes, every flower

relieved by grass blades and green leaves, the
whole devoid of any trace of man and his mud-
dlings in the earth, or his exceeding weakness for
tracing wall-paper patterns, where everything should
be varied, indefinite, and changeful, as the flowers
that bloom and die ; and he would acknowledge
that you had indeed caught the true meaning of
nature in her disposition of vegetation, without
sacrificing one jot of anything in your garden,
but, on the other hand, adding the highest
beauty to spots hitherto devoid of the slightest
interest.

It is not only to places in which shrubberies, and
plantations, and belts of grass in the rougher parts
of the pleasure-ground, and shady moss-bordered
wood-walks occur that these remarks apply.
The suburban gardener, with his single fringe of
planting, may do likewise, to some extent, with
the best taste. He may have the Solomon's Seal
arching forth from a shady recess behind a tuft of the
sweet-scented Narcissus, while in every case he can
make preparations for wild fringes of strong and
hardy spring flowers. In front of a shrubbery with
a sunny aspect is the best of all places for a cheerful
display in early spring, as the shelter and warmth
combined make them open forth in all their glory

under a spring sun, and they cannot be cut off by harsh winds as when exposed in the open garden. What has already been stated is, I hope, sufficient to hint to everybody the kind of place that may be used for their culture. Wild and semi-wild places, rough banks in or near the pleasure-ground or flower-garden, such spots as perhaps at present contain nothing but weeds, or any naturally rough or unused spot about a garden—such are the places I recommend. It is true there are thousands of places without these, and where every inch of the lawn must be mown ; but even on such the Snowdrop may be enjoyed in early spring, for its leaves die down, or at all events ripen sufficiently before there is any occasion to mow the grass.

I have spoken of the Buttercups ; let us next see what may be done with the order to which they belong. It embraces many subjects widely diverse in aspect from these burnished ornaments of northern meadows and mountains. The first thing I should take from it to perennially embellish the wild wood is the sweet-scented Virgin's Bower (Clematis Flammula), a native of the south of Europe, but as hardy and free in all parts of Britain as the common Hawthorn. And as the Hawthorn sweetens the breath of early summer, so will this add fragrance to the

autumnal months. It is never to be seen half so
beautiful as when crawling over some old rockwork
or decayed stumps of trees ; it is excellent for
gathering in wreaths for use along with other flowers
in autumn ; and if its profuse masses of white
bloom do not attract, its fragrance is sure to do so.
An open glade in a wood, or open spaces on banks
near a wood or shrubbery, would be charming for
it ; while in the garden or pleasure-ground it may
be used as a creeper over old stumps, trellising, or
the like.   C. campaniflora, with flowers like a campanula, and of a pale purplish hue, and the beautiful white Clematis montana grandiflora, a native
of Nepaul, are almost equally beautiful, and many
others of the family are worthy of naturalization.

　　The fine new hybrids and varieties (in the way of
C. lanuginosa) will, on good warm sandy soil, spread
over the ground without any support or training,
and in the most luxuriant way.   In making mixed
borders, rockwork, fringes of plantation, or anything
of the kind, we must not be confined by any rules
except those of the judgment, and must draw from
all sorts of stores ; therefore these new varieties of
Clematis should not be overlooked, and if one were
making a bold rockwork, a grand use might be
made of them for dressing precipitous points with

richest colour and noblest flowers, putting the roots in a position where they could descend at pleasure into a rich and deep vein of good earth. The warmth of the recumbent position on the stone, and the shelter, could not fail to make them feel at home, and I can imagine nothing more effective than a sheet of these falling over the face of such large stones as those in the rockwork at Chatsworth, and a few other gardens where large things in this way have been attempted. The beauty displayed by these large varieties of Clematis when planted in a deep light soil is only to be realized by those who have seen it.

Next we come to the Wind Flowers, or Anemones, and here we must pause to select, for a more attractive class of hardy flowers does not beautify any northern clime. Have you a bit of bare, stony ground, slightly shaded perhaps? If so, the beautiful downy white and yellow Anemones of the Alps (A. alpina and A. sulphurea) may be grown there. Any kind of wood or shrubbery which you wish to embellish with the choicest vernal beauty? Then select Anemone blanda, a small but lovely blue kind ; place it in open bare spots to begin with, as it is very dwarf, and it will at Christmas, and from that time onward through the spring, open flowers

as large as a five-shilling piece, and of the deepest
sky blue. The common garden Anemone (A. Coronaria) will not be fastidious, but had better be
placed in open bare places ; and the splendid Anemone fulgens, when it can be spared for the purpose, will prove a most attractive ornament, as it
glows with the most fiery scarlet. It should have
an open spot where the herbage is dwarf. Of other
Anemones, hardy, free, and beautiful enough to be
made wild in our shrubberies, pleasure-grounds, and
wilds, the Japan Anemone (A. japonica), and its
white descendant, A. j. Honorine Jobert, A. trifolia,
and A. sylvestris, are the best of the exotic species.
The Japan Anemone and A. hybrida, and the white
Honorine Jobert, grow so strongly that they will
take care of themselves even among stiff brushwood,
brambles, &c. ; and they are beautifully fitted for
scattering along the low, half-wild margins of shrubberies, &c. The interesting little A. trifolia is not
unlike our own wood Anemone, and will grow in
similar places.

As for the Apennine Anemone, it is simply one
of the loveliest spring flowers of any clime, and
should be in every garden, in the borders, and scattered thinly here and there in woods and shrubberies,
so that it may become " naturalized." The flowers

are freely produced, and of the loveliest blue. It is scarcely a British flower " to the manner born," so to speak, being a native of the south of Europe ; but having strayed into our wilds and plantations occasionally, it is now included in most books on British plants. A. ranunculoides, a doubtful native, found in one or two spots, but not really British, is well worth growing, being very beautiful, and forming tufts of golden yellow.

The beautiful new and large A. angulosa I have seen growing almost as freely as Celandine among shrubs and in half-shady spots, and we all know how readily the old kinds grow on all garden soils of ordinary quality. There are about ten or twelve varieties of the common Hepatica (Anemone Hepatica) grown in British nurseries and gardens, and all the colours of the species should be represented in every collection of spring flowers.

Many will doubtless remember with pleasure the prettily-buttoned white flowers of the Fair Maids of France (Ranunculus aconitifolius fl. pl.,) and in a half-shady rich border it is a beautiful and first-class plant ; but I am disposed to think more of the double varieties of the British Ranunculuses, because of their greater hardiness and vigour. Weed as is the common R. acris, its double variety, with the perfectly formed and polished golden buttons, is a charming hardy plant, flowering profusely, and not of a very transient character; the flowers are even useful for cutting. Good also are R. repens fl. pl. and R. bulbosus fl. pl. R. montanus is a pretty little species, better suited for rockwork, stony ground, or a spot where it would be safe from injury; it is very dwarf and neat, and the flowers comparatively large. Quite distinct from all these, and of chastest beauty when well grown, is R. amplexicaulis, with flowers of pure white, and simple leaves of a dark glaucous green and flowing graceful outline; a hardy and charming plant on almost any soil. It is, indeed, a beautiful and distinct plant, and generally speaking so rare, that had I not seen it selling in the Nottingham market for a few pence per tuft (!) some months ago, I should not mention it here, for usually it is rare even in botanic gardens, and I was much surprised to see it selling like Musk-Plant or Bachelor's-buttons.

There is, however, a handsome double variety of our fine wild Marsh-marigold sold rather plentifully in London during the spring months, and it, unlike the single one, is not so generally known or grown as it ought to be. Of the Globe Flowers (Trollius), the best are T. Napellifolius, T. asiaticus, and the British T. europæus. These are all rich in colour, fragrant, and striking in a remarkable degree.

The Winter Aconite (Eranthis hyemalis,) should be naturalized in every country seat in Britain—it is as easy to do so as to introduce the thistle. It may be placed quite under the branches of deciduous trees, will come up and flower when they are as naked as stones, have its foliage developed before the leaves come on the trees, and be afterwards hidden from sight. Thus masses of this earliest flower may be grown without the slightest sacrifice of space, and only be noticed when bearing a bloom on every little stem. That fine old plant, the Christmas Rose, (Helleborus niger,) likes shade or partial shade better than full exposure, and should be used abundantly, giving it rather snug and warm positions, so that its flowers may be encouraged to open well and fully. Any other kinds of which there was a surplus stock might also be used. And here I might incidentally suggest that every time the borders of hardy plants are dug over, the trimmings and parings of many garden ornaments will do for planting in the woods and wilds.

Of the Monkshoods the less we say the better, perhaps. Some of them are handsome, but all of them virulent poisons ; and, bearing in mind what damage has been done by them from time to time, they are better not used at all. Not so the Delphiniums, which are amongst the most beautiful of all flowers. They are now to be had in such profuse variety that particular kinds need not be named, all being good. A " mixed" packet of seed from any seedsman would afford a number of fine plants. They embrace almost every shade of blue, from the rich dark tone of D. grandiflora to the charming cærulean tints of such as D. Belladonna ; and being usually of a tall and strong type, will make way among long grasses and vigorous weeds, unlike many things for which we have to recommend an open space, or a wood with nothing but a carpet of moss under the trees.

We have thus seen, from examples of three groups, what may be done in the way I propose. I might go through all the other orders in the same way, but as this is done more systematically further on, it is not needful here. I might go from glade to glade and bank to bank, and show how a different aspect of vegetation might be produced in each ; but that will be suggested by the natural orders, by the lists of selections, and, better than all, by a knowledge of the plants themselves. One

of the most delightful phases of the subject is
that of naturalizing alpine and rock plants on ruins
and old walls : there are scores of kinds that not
only thrive on such places, but are to be seen to
greater advantage on them than in any other positions ; but as this is very fully dealt with in an
illustrated chapter in my "Alpine Flowers," I
content myself in the present work with giving a
carefully drawn up list of the best species that will
succeed on ruins and old walls.

By these means it is quite practicable to create
aspects of vegetation along our wood and shrubbery
walks, and in neglected places, superior to any
seen in nature, because we may cull from the flora
of every northern, temperate, and alpine region ;
whereas in nature comparatively few plants exist
wild in a restricted space, while the effect of the
planting which I suggest need be in no sense
inferior in any one spot to that of the sweetest
wild of Nature's own arranging.

It must not be thought that my proposal can
only be carried out in places where there is some
extent of rough pleasure-ground, or some approxi-
mation to what I call half-wild places. Undoubtedly the finest effects may be obtained in
these ; but excellent results may be obtained from

the system in comparatively small villa-gardens, on the fringes of shrubberies, and marginal plantations, open spaces between shrubs, the surface of beds of Rhododendrons, etc. In a word, every shrubbery and plantation surface that is so needlessly and relentlessly dug over by the gardener every winter, may be embellished in the way I propose, as well as wild places. As I have said in " Alpine Flowers," no practice is more general, or more in accordance with ancient custom, than that of digging shrubbery borders, and there is none in the whole course of gardening more profitless or worse. When winter is once come, almost every gardener, although animated with the best intentions, simply prepares to make war upon the roots of everything in his shrubbery border. The generally accepted practice is to trim, and often to mutilate the shrubs, and to dig all over the surface that must be full of feeding roots. Delicate half-rooted shrubs are often disturbed ; herbaceous plants, if at all delicate and not easily recognised, are destroyed ; bulbs are often displaced and injured ; and a sparse depopulated aspect is given to the margins, while the only "improvement" that is effected by the process is the annual darkening of the surface by the upturned earth.

Illustrations of my meaning occur by miles in our London parks in winter. Walk through any of them at that season, and observe the borders round masses of shrubs, choice and otherwise. Instead of finding the earth covered, or nearly covered, with vegetation close to the margin, and each individual developed into something like a respectable specimen of its kind, we find a spread of recently-dug ground, and the plants upon it with an air of having recently suffered from a whirlwind, or some calamity that necessitated the removal of mutilated branches. Rough-pruners precede the diggers, and bravely trim in the shrubs for them, so that nothing may be in the way; and then come the diggers, who sweep along from margin to back, plunging deeply round and about plants, shrubs, or trees. The first shower that occurs after this digging exposes a whole network of torn-up roots. There is no relief to the spectacle; the same thing occurs everywhere—in a London botanic garden as well as in our large West-end parks; and year after year is the process repeated.

While such is the case, it will be impossible to have an agreeable or interesting margin to a shrubbery; albeit the importance of the edge, as

D 2

compared to the hidden parts, is pretty much as that of the face to the back of a mirror. Of course all the labour required to produce this happy result is worse than thrown away, as the shrubberies would do better if let alone, and merely surface-cleaned now and then ; but by utilizing the power thus wasted, we might highly beautify the positions that now present so objectionable an aspect.

If we resolve that no annual manuring or digging is to be permitted, nobody will grudge a thorough preparation at first. Then the planting should be so arranged as to defeat the digger. To graduate the vegetation from the taller subjects behind to the very margin of the grass is of much importance, and this could only be done thoroughly by the greater use of permanent evergreen and very dwarf subjects. Happily, there is quite enough of these to be had suitable for every soil. Light, moist, peaty, or sandy soils, where such things as the sweet-scented Daphne Cneorum would spread forth its dwarf cushions, would be somewhat more desirable than say, a stiff clay ; but for every position suitable plants might be found. Look, for example, at what we could do with the dwarf-green Iberises, Helianthemums, Aubrietias, Arabises, Alyssums, dwarf shrubs, and little conifers

like the creeping Cedar (Juniperus squamata), and
the Tamarix-leaved Juniper! All these are green,
and would spread out into dense wide cushions,
covering the margin, rising but little above the
grass, and helping to cut off the formal line which
usually divides margin and border. Behind them
we might use very dwarf shrubs, deciduous or evergreen, in endless variety ; and of course the margin
should be varied also.

In one spot we might have a wide-spreading tuft
of the prostrate Savin pushing its graceful evergreen branchlets out over the grass ; in another the
dwarf little Cotoneasters might be allowed to form
the front rank, relieved in their turn by pegged-
down Roses ; and so on without end. Herbaceous
plants, that die down in winter and leave the
ground bare afterwards, should not be assigned any
important position near the front. Evergreen
Alpine plants and shrubs, as before remarked, are
perfectly suitable here ; but the true herbaceous
type, and the larger bulbs, like Lilies, should be
"stolen in" between spreading shrubs rather than
allowed to monopolize the ground. By so placing
them, we should not only secure a far more satisfactory general effect, but highly improve the
aspect of the herbaceous plants themselves. Of

course, to carry out such planting properly, a little
more time at first and a great deal more taste than
are now employed would be required ; but what a
difference in the result ! In the kind of borders
I advocate, nearly all the trouble would be over
with the first planting, and labour and skill could be
successively devoted to other parts of the grounds.
All that the covered borders would require, would be
an occasional weeding or thinning, &c., and perhaps
in the case of the more select spots, a little top-
dressing with fine soil. Here and there, between
and amongst the plants, such things as Forget-me-
nots and Violets, Snowdrops and Primroses, might
be scattered about, so as to lend the borders a
floral interest even at the dullest seasons; and
thus we should be delivered from digging and
dreariness, and see our ugly borders alive with
exquisite plants. The chief rule should be—never
show the naked earth : carpet or clothe it with
dwarf subjects, and then allow the taller ones to
rise in their own wild way through the turf or
spray. It need hardly be said that this argument
against the digging applies to two or three beds of
shrubs, and places where the "shrubbery" is little
larger than the dining-room, as much as to the
large country seat, public park, or botanic garden.

It would require a long list to enumerate the many unattractive places that may be beautified by the adoption of this system of naturalization. Take for example a common ditch shaded with trees. There would be no difficulty in enumerating many plants that would thrive better in such a position, with a little clearing and preparation, than we have ever seen them do in any position they now occupy in gardens. It would in fact be a perfect paradise for such plants as Trillium grandiflorum and other inhabitants of dense woods. My friend Dr. Hudson, of Dublin, has converted an old ditch of this kind bordering his place at Merrion into a very agreeable walk, by simply putting a foot or so of coal-ashes and lime-rubbish into it so as to form a dry walk ; and the banks of this shady, narrow alley, he will convert into " mixed borders" of the most charming kind, by selecting plants that love, and thrive in, shady sheltered spots, and by so arranging them that no two parts of the scene shall present the same aspect of vegetation.

I will next enumerate, and indicate the best positions for, the plants suitable for the system.

PART II.

AN ENUMERATION

OF

HARDY EXOTIC PLANTS,

SUITABLE FOR

Naturalization in our Woods, Semi-wild Places, Shrubberies, etc.,

WITH THE

NATIVE COUNTRY, GENERAL CHARACTER, HEIGHT, COLOUR,
TIME OF FLOWERING, MODE OF PROPAGATING, AND
THE POSITIONS MOST SUITABLE FOR EACH.

HARDY EXOTIC PLANTS

FOR NATURALIZATION.

THE BUTTERCUP FAMILY.

Hare-bell Virgin's Bower. *Clematis campaniflora.*
Native country: S. Europe. *Habit:* a climber. *Height:*
6 to 10 feet. *Colour of flower:* purplish. *Time of flowering:* summer. *Manner of propagation:* by seed, as in
all the kinds, to be sown as soon as gathered, division, or
layers.—*Suitable positions:* copses, banks, old stumps,
hedgerows, &c. in ordinary soil.

American Traveller's Joy. *Clematis Viorna.* North
America. Climber; 8 to 10 feet; purple; summer and
early autumn; seed, division, or layers.—Thin low copses,
open sunny banks, rootwork, hedgerows, etc.

Vine-bower Clematis. *Clematis Viticella.* South
Europe. Climber; 10 to 16 feet; blue or purple;
summer and early autumn; seed or layers.—Fringes of
woods, copses, hedgebanks; through wild or semi-wild
shrubby vegetation on high banks, tall old stumps, or
high rootwork.

Sweet-scented Virgin's Bower. *Clematis Flammula.*
Southern Europe. Climber; 10 to 30 feet; white;
autumn; seed or layers.—Excellent for almost every use
to which a hardy climber may be put, and in the semi-

wild state for banks, stumps, chalk-pits, hedges, copses, and even for planting in masses in grassy places.

Richer sheets of noble bloom are not to be seen in the open air in any northern clime than those produced by the new hybrid clematises raised by Jackman of Woking and others. They are capable of beautifying any position, and seem to conform to almost any mode of culture or training—pegged down, trained up on stakes, or nailed against walls ; but there is certainly no spot which suits them so well as the face of a large rock, natural or artificial. Planted in deep good soil, above and behind such an object, the shoots will fall over the face of the rock in vigorous matted tufts, and in due season become so densely covered with flowers as to resemble a truly imperial robe of purple. They may also be planted so as to fall over the side-walls of rustic bridges either over walks or streams, and may be allowed to run over the face of bare sunny banks, where they would produce a magnificent effect. The variety best known at present is Jackman's (*Clematis Jackmani*); but there are many other kinds.

Meadow Rues. *Thalictrums.* This large and well-marked family is of somewhat too coarse and weedy a nature for garden culture ; but, being possessed of a very vigorous habit, and being also distinct in aspect, it is precisely one of those that are suitable for planting here and there in the wildest and roughest parts of our planta-tions. Of the rather numerous kinds of these grown in our botanic gardens, the most ornamental are the plumy Meadow-rue and the fetid Meadow-rue : as these are capable of producing distinct and desirable effects, I will speak of them separately. As regards all the other species likely to be met with in gardens, and including also any plants of the plumy Meadow-rue, they may be planted among any coarse herbaceous vegetation. For the most part they attain a height of three or four feet, and are as easily propagated by division as the common balm.

Fetid Meadow Rue. *Thalictrum fœtidum.* Europe. Herbaceous perennial; 9 inches to one and a half feet high; brownish; summer; division or seed.—A plant not worthy of cultivation on account of its flowers; but having very gracefully cut leaves, very like those of our own Lesser Meadow-rue—and resembling, when grown on established plants, those of the Stove Maiden-hair fern (*Adiantum cuneatum*), it deserves to be grown, as does also the Lesser Meadow-rue, for the beauty of its leaves. It is, like that plant, hardy enough to grow in almost any soil or position, but will be seen to greatest advantage on open spots or banks with a dwarf vegetation of late spring and early summer flowers. In such places tufts of it ought to look as well as plants of the Maiden-hair fern do among conservatory flowers. It is, however, only just to the British Thalictrum minus to say that it produces a very similar effect and quite as good, so that anybody possessing it need not seek our present subject.

Plumy Meadow Rue. *Thalictrum aquilegifolium.* Middle and Southern Europe. Herbaceous perennial; 3 to 4 feet; whitish rose or purplish; summer; division. —Will grow in almost any soil or position, but prefers a somewhat humid spot. The variety with purplish instead of yellow stamens is a pretty one, and both are well suited for a position near wood walks.

Alpine Wind-Flower. *Anemone alpina.* Alps. Herbaceous perennial; 4 to 20 inches; white and purplish on the outside of the petals; summer; seed and division.—On grassy banks, in unmown parts of the pleasure-grounds or open spots in woods, in which it ought to attain as great perfection as it does in sub-alpine meadows.

Apennine Wind-Flower. *Anemone apennina.* Europe. Tuber; 3 to 9 inches; blue; spring; division. —Rocks, stony places, in exposed positions, and also in bare shady or half-shady places, in groves, and by the side of avenues and wood walks. It may, in fact, be grown with success wherever the common wood-anemone thrives.

Poppy Wind-Flower. *Anemone Coronaria.* Levant. Tuber; 6 to 12 inches; striped; spring; seed and division.—Open sunny places, fringes of shrubberies, banks, etc., where there is a dwarf vegetation.

Japanese Wind-Flower. *Anemone japonica.* Japan. Herbaceous perennial; 2 feet; reddish; autumn; division. —Woods, copses, brakes, amongst masses of Cotoneaster and other prostrate shrubs, margins of shrubberies, in fact in almost any position and soil.

White Japanese Wind-flower. *Anemone japonica var. Honorine Jobert.* Garden variety. Herbaceous perennial; 2 feet; white; autumn; division.—Similar positions to those for the preceding, than which it is even a finer plant.

Crowfoot Wind-Flower. *Anemone ranunculoides.*

spring; seeds or division. — Margins of shrubberies, copses, and by wood-walks associated occasionally with the Alpine Anemone, and the finer Crowfoots.

Hepatica. *Anemone Hepatica.* Europe. Herbaceous perennial; 3 or 4 inches; various colours; spring; division.—A native of mountain woods, this thrives very well in bare places, in shady or open woods and shrubberies; also in rocky places, the chief care required being to plant it where its beauties may be seen.

Three-leaved Wind-Flower. *Anemone trifolia.* France. Tuber; 6 to 9 inches; white; spring or early summer; seed or division.—Suitable for the same uses as the wood and Apennine wind-flowers, and for association with them.

Aconite-leaved Crowfoot. *Ranunculus aconitifolius.* Europe. Herbaceous perennial; 1 to 1½ feet; white; early summer; seed or division.—Similar positions to those for the Alpine wind-flower.—It is often found wild in great luxuriance in rather low meadows under conditions nearly like those enjoyed by our own meadows.

Large Double Crowfoot. *Ranunculus " bullatus."* *fl. pl.* Garden variety. Herbaceous perennial; ½ foot; yellow; summer; division.—A handsome double variety, which will thrive well by wild wood-walks, and in the rougher parts of the pleasure-ground, where the vegetation is dwarf.

Stem-clasping Crowfoot. *Ranunculus amplexicaulis.* Pyrenees. Herbaceous perennial; 1 foot; white; spring and early summer; division. — A lovely subject for naturalization in open rocky spots, where there is a moist free soil. Near the margins of a mountain bog would suit it well, though it is yet so scarce that all the supply is required for the select rockwork. It might also be tried with success in an open bare spot, amidst vegetation, not rising above six inches high.

Glacier Crowfoot. *Ranunculus glacialis.* Lapland. 6 inches to a foot; white ; summer; seed or division.—Worthy of attention where there is stony and cold mountain ground, elevated bogs, and the like. In such places one might be proud of having naturalized such a high Alpine plant.

Mountain Crowfoot. *Ranunculus montanus.* Austria. Herbaceous perennial ; 6 to 9 inches ; yellow ; summer ; seed or division.—Bare spots, or where the vegetation consists of such plants as the spurrey or the shortest grasses, and where the soil is somewhat moist and free.

Spiked Crowfoot. *Ranunculus spicatus.* Algiers. Herbaceous perennial ; 1 to 1½ feet ; yellow ; spring and summer ; division.—Excellent for association with the Snowdrop Anemone, and other choice plants, reaching a height of something over one foot, by wood-walks, in rather open sunny spots.

Altaian Globe-flower. *Trollius altaicus.* Altai mountains. Herbaceous perennial ; 2 feet ; yellow ; summer ; division.— By shrubbery walks in unmown places, amidst rather strong-growing, herbaceous vegetation, and also in open grassy glades in woods.

Napellus-leaved Globe-flower. *Trollius napellifolius.* Europe. Herbaceous perennial ; 2 or 3 feet ; yellow ; early summer ; division. — A noble plant, useful for positions similar to those for the preceding kind, and for association with it.

Common Winter Aconite. *Eranthis hyemalis.*

Europe. Tuber; 4 inches; yellow; winter; division.—
Bare places in woods or copses, shady or sunny
banks, and also under isolated trees, of which the
branches rest on the grass of the lawn or pleasure-
ground.

Christmas Rose. *Helleborus niger.* Europe. 1 foot;
white; winter; division.—A well known plant, which
will be seen to greater advantage on sunny, yet sheltered
grassy banks, than on the margins of shrubberies, if any
choice as to position may be made.

Olympian Hellebore. *Helleborus olympicus.* India.
Evergreen perennial; 2 feet; green; winter; division.—
Grassy, sunny banks, like preceding, and also for the
margins of shrubberies, and for broken ground.

Alpine Columbine. *Aquilegia alpina.* Switzerland.
Herbaceous perennial; 9 to 12 inches; blue; early
summer; seed or division. — Well worthy of naturalization in cool, moist, sandy soil, in stony places
near cascades, and on similar places in elevated positions.

Canadian Columbine. *Aquilegia canadensis.* N.
America. Herbaceous perennial; 1 to 2 feet; red and
orange; spring and early summer; seed or division.—
Sandy soils in rather open spots, amidst thin grass and
not very rampant herbaceous plants.

Sky-blue Columbine. *Aquilegia cærulea.* N. America. Herbaceous perennial; 1 to 2 feet; blue and
white; summer; seed or division.—Excellent in very
sandy, well-drained soil in an open position among herbs
not too vigorous in habit, and not reaching much over
one foot high.

E

Perennial Larkspur. *Delphinium.* Gardens are now enriched by a multitude of beautiful and vigorous varieties of perennial Delphiniums, and new kinds of the most delicate and attractive appearance are annually raised. Growing from 2 to nearly 6 feet high, perfectly hardy and thriving in ordinary soil, any of these plants which may be spared from the garden should be planted out in half-wild places amidst herbaceous vegetation. They would be particularly appropriate in wide open spaces in woods or near wood-walks, associated with Pæonies, Asters, and plants of like stature.

Variegated Monkshood. *Aconitum variegatum.* Southern Europe. Herbaceous perennial; 4 to 6 feet; purple and white; summer; division.—Makes noble tufts in positions recommended for the Delphiniums, and is suited for association with the most vigorous and showy herbs. Other kinds of Monkshood are ornamental, but they are all so frightfully poisonous that even this one is perhaps better avoided.

White-flowered Pæony. *Pæonia albiflora.* Siberia. Herbaceous perennial; 2 to 3 feet; white; early summer; seed or division.—Margins of shrubberies, wood-walks, etc. From this species have sprung many of the noble varieties of Pæony which are now in cultivation.

Officinal Pæony. *Pæonia officinalis and vars.* Europe. Herbaceous perennial; 3 to 4 feet; red; division.—Rough rocky places, banks, and edges of woods and copses.

Fine-leaved Pæony. *Pæonia tenuifolia.* Siberia. Herbaceous perennial; 1 to 1½ feet; red; early summer: division.—Rough rockwork, margins of low shrubberies, rocky places, banks, and glades.

THE BARBERRY FAMILY.

Pinnate Barren Wort. *Epimedium pinnatum.* Persia. Herbaceous perennial; 9 to 30 inches; yellow; spring; division.—Warm half shady spots on the margins of shrubberies, or beds of American plants, or naturalized in copses, in moist, peat, or vegetable soil.

Alpine Barren Wort. *Epimedium alpinum.* Southern Europe. Herbaceous perennial; 9 to 12 inches; purplish; spring; division.—Same positions as for the preceding, but amid dwarfer vegetation.

THE DUCK'S FOOT FAMILY.

May Apple. *Podophyllum peltatum.* North America. Herbaceous perennial; 6 to 9 inches; white; early summer; seed or division.—Grows quite vigorously in half-shady spots on the margins of beds of Rhododendrons, &c., in peat soil. An interesting plant, yielding the now popular medicine, Podophyllin.

THE WATER LILY FAMILY.

Large Yellow Water Lily. *Nuphar advena.* North America. Aquatic; yellow; summer; division.—A noble plant for the margins of ornamental water, associated with our beautiful British Water-Lily.

THE POPPY FAMILY.

Saffron-coloured Poppy. *Papaver croceum.* Altai Mountains. Herbaceous perennial; 1 to 1½ feet; saffron; early summer; seed or division.—Rocky ground in moist districts, in rather moist sandy soil.

Great Scarlet Poppy. *Papaver bracteatum.* Siberia. Herbaceous perennial; 3 to 5 feet; red; early summer; seed or division.—Open spots in woods ; a splendid plant in almost any position, growing well in the worst soils.

Naked-stemmed Poppy. *Papaver nudicaule.* Siberia. Herbaceous perennial; 9 inches to 1½ feet; yellow; summer ; seed.— Open rocky ground in moist sandy soil —a very handsome plant.

Oriental Poppy. *Papaver orientale.* Eastern Europe. Herbaceous perennial; 2½ to 4 feet ; red ; early summer ; seed or division.—Nearly allied to P. bracteatum, and also a magnificent plant for naturalization.

Opium Poppy. *Papaver somniferum.* Southern Europe. Annual ; 3 to 4 feet ; various ; summer ; seed.— The varieties of this are showy in sunny spots in open parts of copses and woods, growing in any soil.

Canadian Blood Root. *Sanguinaria canadensis.* North America. Tuber ; ½ foot; white; spring or early summer ; division.—Bare places in woods and copses.

Cordate Macleya. *Macleya cordata.* China. Herbaceous perennial; 4 to 6 feet; brownish; summer; division.—As single specimens by half-wild pleasure-ground walks, growing best in deep sandy loam.

Californian Eschscholtzia. *Eschscholtzia californica.* North America. Annual; 1 foot ; yellow ; all summer ; seed.—Any place rather open, and amidst rather dwarf vegetation. It comes up self-sown year after year.

Californian Platystemon. *Platystemon californicum.* N. America. Annual; 6 to 12 inches ; yellow ; summer ; seed.—On very bare and open spots, where there is a spare and minute vegetation, in light soil.

THE FUMITORY FAMILY.

Plumy Dielytra. *Dielytra eximia.* North America. Herbaceous perennial; 1 foot; flesh-colour; summer; division.—Amid rock plants, on the fringes of very low shrubberies, on bare banks, or in open rocky spots in any soil.

Showy Dielytra. *Dielytra spectabilis.* Siberia. Herbaceous perennial; 1 to 2 feet; fine rose; early summer; seed or division.—Fringes of woods, and shrubberies, and all like positions, in almost any soil.

Yellow Corydalis. *Corydalis lutea.* Southern Europe. Perennial; 1 to 2 feet; yellow; all summer; seed.—Loves stony banks and old ruins, and often establishes itself on very high walls not in a ruinous condition. It is quite at home on very ugly common-place rockworks on which little else will thrive.

THE CRUCIFER FAMILY.

Annual Stock. *Matthiola annua.* Southern Europe. Annual; 1 to 2 feet; red; all summer; seed.—Any bare open spots in woods or copses, or on stony banks, or old ruins.

Window Stock. *Matthiola fenestralis.* Southern Europe. Biennial; 1 foot; purple; late in summer; seed.—Sunny margins of shrubberies, and in deep good soil.

Two-horned Stock. *Matthiola bicornis.* Greece. Annual; 6 to 12 inches; bright rosy purple; summer; seed.—Warm and bare, or stony open ground amidst low vegetation; it is deliciously scented.

Emperor Stock. *Matthiola semperflorens.* Evergreen perennial; 1 to 1½ feet; summer; seed.—Rocky places, banks, ruins, fringes of shrubberies, etc.

Night-scented Stock. *Matthiola tristis.* Southern Europe. Biennial; 1 to 1½ feet; brown; summer; seed.—May be established on the sunny sides of old ruins and walls, in old chalk-pits, etc.

Alpine Wallflower. *Erysimum ochroleucum.* Northern Europe. Evergreen perennial; ¼ foot; yellow; early summer; cuttings, seed, or division.—Bare or stony earth in moist soil, and in a fully exposed position.

Garden Wallflower. *Cheiranthus Cheiri.* Europe. Evergreen perennial; 2 feet; rich brown and yellow; early summer; seed or cuttings.—This, as everybody knows, is quite at home on old walls, on many of which it is abundantly naturalized.

White Wall-cress. *Arabis albida.* Caucasus. Evergreen perennial; ¾ foot; white spring; seed, division, or cuttings.—Anywhere amid vegetation not over 1 foot high Should be used abundantly in stony places and ruins.

Sand Wall-cress. *Arabis arenosa.* Europe. Annual; ½ foot; purplish; seed.—On mossy old ruins or walls, or on ground fully exposed, and where the vegetation is not much more than 6 inches high.

Biennial Honesty. *Lunaria biennis.* Germany. Biennial; 2 to 3 feet; purplish; summer; seed.—On warm chalky banks, or slopes, or indeed in almost any position in woods, and by walks in half-wild spots.

Perennial Honesty. *Lunaria rediviva.* Europe. Herbaceous perennial; 2 to 3 feet; purple; early

summer; seed or division.—Margins of shrubberies, beds
of American plants, and copses.

Purple Aubrietia. *Aubrietia deltoidea, and vars.*
Southern Europe. Evergreen perennial; 4 inches;
purple; spring and early summer; seed, division, or
cuttings.—Anywhere amidst very dwarf vegetation: on
rocky or bare ground, banks or slopes, it should be
planted in profusion.

Alpine Mad-wort. *Alyssum alpestre.* Southern
Europe. Evergreen perennial; 4 inches; yellow;
summer; seed or cuttings.—A very neat little plant to
establish on bare rocky upland ground, or on ruins.

Rock Mad-wort. *Alyssum saxatile.* Russia. Evergreen perennial: 12 to 18 inches; yellow; spring and
early summer; seed, cuttings, or division.—Should be
abundantly planted on bare and rocky ground, on banks
and slopes, associated with all showy alpine plants like
the white Arabis and purple Aubrietia.

Sweet Alyssum. *Alyssum maritimum.* Southern
Europe. Annual; 1 foot; white; summer; seed.—Stony
or bare ground where the vegetation is dwarf and sparse.

Mountain Mad-wort. *Alyssum montanum.* Germany.
Evergreen perennial; 3 to 6 inches; yellow; summer;
seed, cuttings, or division.—On rocky ground that is
rather sandy and not too wet, or on bare banks.

Sea-green Whitlow Grass. *Draba aizoides.* Europe.
Evergreen perennial; 3 inches; yellow; spring; seed or
division.—Old ruins, walls, and rocks, on which there is
a very minute vegetation. It is naturalized in one or
two places in England.

Stemless Violet-Cress. *Ionopsidium acaule.* Portugal.

Annual; 1 to 2 inches; lilac; summer; seed.—Only where the vegetation is no larger than mosses, on moist sandy slopes, rocky or bare ground.

Broad-leaved Bastard-Cress. *Thlaspi latifolium.* Herbaceous perennial; 1 foot; white; early spring; seed or division.—Positions similar to those recommended for the White Arabis. Good also for low fringes of shrubberies or bare parts of copses, associated with early flowers.

Coris-leaved Candy Tuft. *Iberis corifolia.* Southern Europe. Evergreen perennial; 6 to 9 inches; white; early summer; seed or cuttings.—Rough rockwork, stony places, or bare banks, fully exposed. The smallest good evergreen Candy Tuft. It ought to be placed amidst very dwarf vegetation.

Gibraltar Candy Tuft. *Iberis gibraltarica.* Spain. Evergreen perennial; 1 foot and over; pinkish; spring and early summer; seed or cuttings.—Warm spots on banks or rocky places in the milder parts of the country.

Garrex's Candy Tuft. *Iberis Garrexiana.* Pyrenees. Evergreen perennial; 6 to 9 inches; white; summer; seed or cuttings.—Same positions as for the preceding kind.

Crown Candy Tuft. *Iberis Coronaria.* Southern Europe. Annual; 1 foot, white; summer; seed.— Open spots in any aspect in ordinary soil.

Rock Candy Tuft. *Iberis saxatilis.* Southern Europe. Evergreen perennial; 6 to 12 inches; white; early summer; seed or cuttings.—Rocks, banks, slopes, margins of shrubberies or woods; best in open spots.

Tenore's Candy Tuft. *Iberis Tenoreana.* Naples. Probably biennial; 6 inches; white, changing to pale purple; summer; seed.—Similar positions to the pre-

ceding, but warmer and always in light, sandy soil, or
thoroughly well drained sandy loam.

Purple Candy Tuft. *Iberis umbellata.* Southern
Europe. Annual; 1 foot; purple; summer; seed.—
Bare open spots in woods or on slopes, in any soil.

Common Rocket. *Hesperis matronalis.* Europe.
Biennial; 1 to 3 feet; various; summer; seed.—In
shrubberies; the single variety will sow itself very
freely. The double ones would probably "die out."

Virginian Stock. *Malcolmia maritima.* Southern
Europe. Annual; 6 to 12 inches; lilac purple; summer;
seed.—Pretty in any spots where it would not be overrun
by grass, etc.

Peroffski's Erysimum. *Erysimum Peroffskianum.*
Palestine. Annual; 1 to 2 feet; deep orange; summer;
seed.—Rather bare spaces in copses, and on banks, in
ordinary soil, "sows itself."

Coris-leaved Æthionema. *Æthionema coridifolium.*
Southern Europe. Evergreen perennial; 6 inches; lilac
rose; summer; seed or division.—Exposed rocks amidst
dwarf vegetation, or on bare banks.

Heart-leaved Seakale. *Crambe cordifolia.* Caucasus.
Tuber; 4 to 6 feet; white; summer; seed or division.—
Grassy spaces beside wood and pleasure-ground walks.
Isolated plants in rich ground are most effective.

Arabis-like Heliophila. *Heliophila araboides.*
South Africa. Annual; 6 to 9 inches; blue; summer;
seed.—Sunny banks and rocky ground, with sparse low
vegetation, in sandy soil.

THE BASTARD HEMP FAMILY.

Bastard Hemp. *Datisca cannabina.* Southern Europe. Herbaceous perennial; 4 to 6 feet; yellowish; summer; seed or division.—Open grassy spaces by wood-walks, and in spots where its graceful habit may be seen. Male and female plants should be planted together, as the female, laden with fruit, is the more graceful of the two. The male is the one commonly seen in England, but both sexes may be had by raising the plants from seeds.

THE CAPER FAMILY.

Common Caper. *Capparis spinosa.* Southern Europe. Deciduous shrub; 3 to 4 feet; white; summer; seed or cuttings.—I believe this interesting and most beautiful, as well as useful plant, may be grown on old walls and ruins, in chalk pits, and on the sunny flanks of rockwork, in warmer parts of Southern England much as it is in various countries warmer than ours. It should always be placed in as warm and sunny a position as possible, and would be best if arranged so that it should project from the face of the sunniest and warmest part of the wall or ruin on which it is placed.

THE ROCK ROSE FAMILY.

Gum Cistus. *Cistus ladaniferus.* Spain. Evergreen shrub; 2 to 3 feet; white; summer; seed or cuttings.— Rocky ground, stony banks, or almost anywhere in a somewhat dry soil.

There are many others of this family that may be used in like positions. Even more valuable for rocky ground,

slopes, and margins of shrubberies are the numerous kinds
of Helianthemums. They flourish to greatest perfection
in chalky or warm soils.

THE VIOLET FAMILY.

Two-flowered Violet. *Viola biflora.* Europe. Herbaceous perennial; 3 to 6 inches; yellow; spring and
early summer; seed or division.—In moist, rocky, or
stony places, between the stones in rockworks, etc.

Horned Violet. *Viola cornuta.* Pyrenees. Herbaceous perennial; 3 to 6 inches; blue; summer; seed,
division, or cuttings.—Rocks, banks, fringes of low shrubberies, or indeed in almost any position where it may not
be overrun by coarse plants.

Canadian Violet. *Viola canadensis.* North America.
Herbaceous perennial; 4 to 6 inches; white, streaked
with violet; summer; seed or division.—A vigorous kind,
good for running beneath fringes of shrubberies and
woods.

THE SUNDEW FAMILY.

American Grass of Parnassus. *Parnassia asarifolia.*
North America. Herbaceous perennial; 3 to 6 inches;
white; summer; seed or division.—Would be well worth
naturalizing in such moist, boggy spots as our own Grass
of Parnassus delights in.

THE MILKWORT FAMILY.

Bastard-Box. *Polygala Chamæbuxus.* Austria.
Evergreen trailer; 3 to 6 inches; yellowish; early summer; division.—Bare rocky places, in a somewhat moist

peat or fine sandy soil, associated with such dwarf shrubs
as Daphne Cneorum, and Erica carnea.

THE PINK FAMILY.

Tall Gypsophila. *Gypsophila altissima.* Siberia.
Herbaceous perennial; 3 to 5 feet; pinkish; summer;
seed or division.—Banks, rocks, and stony places.

Elegant Gypsophila. *Gysophila elegans.* Eastern
Europe. Annual; 1 to 2 feet; pale rose; summer;
seed.—Same positions as for the preceding kind.

Panicled Gypsophila. *Gypsophila paniculata.* Siberia. Herbaceous perennial; 2 to 4 feet; white; summer; seed or division.—Rough, rocky places and in thin
woods.

Trailing Gypsophila. *Gypsophila prostrata.* Europe.
Herbaceous perennial; 1 foot; pale rose; summer;
seed or division.—Rocks, banks, and heaps of stony
rubbish amid dwarf plants.

Creeping Gypsophila. *Gypsophila repens.* Pyrenees.
Deciduous trailer; 3 to 6 inches; striped; summer; seed
or division.—Same positions as for the preceding.

Steven's Gypsophila. *Gypsophila Steveni.* Iberia.
Herbaceous perennial; 2 to 3 feet; white; summer;
seed or division.—Fringes of shrubberies and in open
spots in thin woods, among strong perennials.

Alpine Pink. *Dianthus alpinus.* Austria. Evergreen perennial; 3 to 4 inches; red; summer; seed or
division.—In peat or very sandy moist soil, on bare and
exposed rocky or very stony spots, amid minute plants.

The beautiful, brilliant, and recently introduced *Dian-*

thus neglectus would succeed in similar positions even more freely than this, as it grows well in common soil.

Carnation. *Dianthus Caryophyllus.* Europe. Evergreen perennial; 1 to 2 feet; various; summer; seed, layers, or pipings.—The Carnation grows in a wild state on walls and ruins, and is worthy of being naturalized in such places : also in very rocky ground or stony banks, in ordinary sandy or gravelly soil.

Common Pink. *Dianthus plumarius.* Europe. Evergreen perennial; ½ foot; white; summer; seed, layers, pipings, or division.—On ruins, walls, banks, and dry rocks, on any of which the plant will prove more enduring than on the level moist ground.

Superb Pink. *Dianthus superbus.* Europe. Evergreen perennial; 1 to 2 feet; pale purple; summer; seed or division.—Sandy moist fields, or open spots in woods : also on similar soil in rocky or stony places.

Rock Tunica. *Tunica Saxifraga.* Europe. Evergreen perennial; 1 foot; pale purple; summer; seed or division.—Walls, ruins, rocks, or dry, bare, and poor soil, on banks or slopes.

Calabrian Soap-wort. *Saponaria calabrica.* Calabria. Annual; 6 to 12 inches; deep rose; all summer; seed.—In half-bare, wild places this popular annual would take care of itself, but it is not nearly so ornamental as the following perennial kind.

Rock Soap-wort. *Saponaria ocymoides.* Alps of Europe. Evergreen trailer; 3 inches; red; early summer; seed, cuttings, or division.—Rocks, banks, stony slopes, or in tufts on the edges of shrubberies.

Lobel's Catchfly. *Silene Armeria.* Southern Europe.

Annual; 1 to 1½ feet; red; late summer and early autumn; seed.—Slopes, banks, and almost anywhere amidst vegetation from 12 to 20 inches high.

Alpine Catchfly. *Silene alpestris.* Austrian Alps. Evergreen perennial; 4 to 6 inches; white; early summer; seed or division.—Well-exposed rocky, stony, or bare places. A beautiful little plant.

Pendulous Catchfly. *Silene pendula.* Sicily. Biennial; 1 foot; red; summer; seed.—Grows in any soil or position, but being rather dwarf, is best in rather bare spots or slopes, among spring and early summer flowers.

Dwarf Catchfly. *Silene Pumilio.* Germany. Evergreen perennial; 4 to 6 inches; rose; summer; seed or division.—When easily obtainable, would be worth trying in moist spots on mountains similar to those inhabited by our own Moss Campion.

Late Catchfly. *Silene Schafta.* Caucasus. Evergreen perennial; 6 to 12 inches; reddish; summer and autumn; seed or division.—Positions similar to those recommended for Silene alpestris.

Rose Campion. *Lychnis coronaria.* Italy. Evergreen perennial; 1 to 2 feet; rosy purple; summer; seed.—Thrives with unusual vigour on dry, warm banks, though it grows well in almost any soil.

Dark-eyed Rock Lychnis. *Viscaria oculata.* Algiers. Annual; 1 foot; rosy purple with dark eye; summer; seed.—In good sandy soil in warm positions.

Balearic Sand-wort. *Arenaria balearica.* Majorca. Evergreen perennial; 1 to 2 inches; white; spring and early summer; seed or division.—This plant crawls over wet or moist rocks, rooting on them, somewhat as a moss would, and forms sheets of starry flowers thereon.

Mountain Sand-wort. *Arenaria montana.* Europe. Evergreen perennial; 4 to 6 inches; white; spring and summer; seed or division.—In good sandy loam, either on bare level ground, or on banks or rocks.

Bieberstein's Mouse-ear Chickweed. *Cerastium Biebersteinii.* Caucasus. Evergreen perennial; 2 to 6 inches; white; summer; seed or division.—Banks, fringes of shrubberies, rough, rocky places, or almost any kind of situation, in ordinary soil.

Large-flowered Cerastium. *Cerastium grandiflorum.* Siberia. Evergreen perennial; 2 to 6 inches; white; summer; seed or division.—Similar positions to the foregoing; is not so common as the preceding or following kind.

Woolly Mouse-ear Chickweed. *Cerastium tomentosum.* Southern Europe. Evergreen perennial; 2 to 6 inches; white; summer; seed or division.—Equally useful and hardy as the two foregoing kinds.

THE FLAX FAMILY.

Alpine Flax. *Linum alpinum.* Austria. Herbaceous perennial; 4 to 9 inches; blue; summer; seed or division.—Open rocky places, or fully exposed spots in sandy soil, associated with the more vigorous alpines.

Narbonne Flax. *Linum narbonnense.* S. France. Herbaceous perennial; 1 to 2 feet; blue; summer; seed or division.—In deep sandy loam amidst medium-sized herbaceous plants, or on banks, rocky or stony ground.

THE MALLOW FAMILY.

Moreni's Mallow. *Malva Morenii.* Italy. Herbaceous perennial; 2 to 3 feet; reddish; late summer; seed or division.—Edges of woods, banks, hedges, etc.

Showy Malope. *Malope trifida, var. grandiflora.* Barbary. Annual; 2 to 4 feet; crimson purple; late summer; seed.—Bare spaces in shrubberies, and also among coarse annual and biennial plants, in any position.

Brilliant Mallow. *Callirhoe involucrata.* N. America. Herbaceous perennial; 1 foot; crimson; summer; seed.—Open places, associated with choice herbaceous and alpine plants, excellent for banks and slopes.

Vine-leaved Kitaibelia. *Kitaibelia vitifolia.* Hungary. Herbaceous perennial; 5 to 8 feet; whitish; late summer; division or seed.—Among the most vigorous herbs; in grassy places near wood-walks.

Fig-leaved Hollyhock. *Althæa ficifolia.* Herbaceous perennial; 4 to 6 feet; yellow or orange; summer; seed or division.—Similar positions to those for the preceding.

Taurian Marsh Mallow. *Althæa taurinensis.* Southern Europe. Herbaceous perennial; 4 to 6 feet; reddish; summer; seed or division.—Amidst tall herbs by the margin of water, grouped with the foregoing.

Downy-leaved Lavatera. *Lavatera Olbia.* France. Half shrubby; 3 to 6 feet; reddish; summer; seed or cuttings.—A warm nook in a chalk-pit, or a position on sunny banks among shrubs, brings out the flowers of this plant in great profusion.

Mallow Rose. *Hibiscus Moscheutos.* North America.

Herbaceous perennial; 2 to 4 feet; purplish; late summer; seed or division.—In rich soil in small open glades near wood walks; to be planted in isolated tufts or small beds, always in warm positions.

Military Hibiscus. *Hibiscus militaris.* North America. Herbaceous perennial; 3 to 5 feet; purple; summer; seed or division.—Similar positions to those for the preceding; a noble plant.

Rose-coloured Hibiscus. *Hibiscus roseus.* France. Herbaceous perennial; 3 to 5 feet; rose; late summer and autumn; seed or division.—Also a very remarkable plant, and suitable for positions like those recommended for the preceding.

THE ST. JOHN'S WORT FAMILY.

Large-flowered St. John's Wort. *Hypericum calycinum.* Southern Europe. Evergreen shrub; 1 to 3 feet; yellow; summer; seed or division.—This well-known plant is, it need hardly be said, capable of enduring any hardship, or of embellishing any position in the wilderness. Any other members of the family that may be admired may be grown in copses and on the fringes of woods in any kind of soil.

THE GERANIUM FAMILY.

Iberian Crane's Bill. *Geranium Ibericum.* Levant. Herbaceous perennial; 1 to 2 feet; violet; summer; seed or division.—Woods, copses, shrubberies, by wood walks, and in open grassy glades.

Lambert's Crane's Bill. *Geranium Lamberti.* Nepaul. Herbaceous perennial; 1 to 2 feet; reddish; summer; seed or division.—Similar positions to those for the preceding, but not so very hardy.

Striped Crane's Bill. *Geranium striatum.* Italy. Herbaceous perennial; 1 to 2 feet; striped; summer; seed or division.—Fringes of shrubberies and low banks.

Wallich's Crane's Bill. *Geranium Wallichianum.* Nepaul. Herbaceous perennial; 1 foot; mauve with purple veins; summer; seed or division. — Similar positions to those for the preceding.

Manescavi's Heron's-Bill. *Erodium Manescavi.* Evergreen perennial; 6 inches to 2 feet; reddish; summer; seed or division.—Margins of shrubberies, or rocky or stony ground, banks, or by open wood walks.

THE INDIAN CRESS FAMILY.

Five-leaved Indian Cress. *Tropæolum pentaphyllum.* Southern America. Trailer; yellow; summer; division or cuttings.—Banks, copses, or any position where it may trail over shrubs, &c., in light soil.

Showy Indian Cress. *Tropæolum speciosum.* Chili. Trailer; red and yellow; summer; division or cuttings.— Against terrace walls, among shrubs, and on slopes, on banks, or bushy rockwork near the hardy fernery; in deep, rich, and light soil. In such positions it is a brilliant plant well worth any trouble to establish.

THE WOOD SORREL FAMILY.

Bowie's Wood Sorrel. *Oxalis Bowieana.* Cape of Good Hope. Bulb; 6 inches; scarlet; spring and summer; division.—In rocky, bare, and sunny places, in light dry loam, or very sandy soil.

Free-flowering Wood Sorrel. *Oxalis floribunda.* Brazil. Evergreen perennial; 9 to 18 inches; red; summer; seed or division.—Among dwarf Alpine plants, in almost any soil or position.

THE BEAN CAPER FAMILY.

Greater Honey Flower. *Melianthus major.* Cape of Good Hope. Herbaceous perennial; 4 to 6 feet; brownish; summer; seed, cuttings, or division.—This elegant-leaved plant will be found to thrive well on slightly elevated banks, in the south of England, in well-drained loam. It may be cut down in winter, but will come up the following season.

THE PEA FAMILY.

Faba-like Thermopsis. *Thermopsis fabacea.* Siberia. Herbaceous perennial; 2 to 3 feet; yellow; summer; seed or division.—Among strong herbaceous plants, by wood walks, on the margins of woods, or in open spots in shrubberies or pleasure-grounds.

Blue False Indigo. *Baptisia australis.* Carolina. Herbaceous perennial, 2 to 3 feet; blue; midsummer; seed or division.—Woods, copses, banks, among low shrubs and stout herbs in any kind of soil.

Cluster-flowered Cytisus. *Cytisus capitatus.* Austria. Shrub; 2 to 3 feet; yellow; summer; seed or cuttings.—In positions similar to the preceding.

Winged Genista. *Genista sagittalis.* Alps of Europe. Small shrub; 6 to 12 inches; yellow; summer; seed or division.—In grassy open places and on banks; also on rocks or slopes.

Shrubby Restharrow. *Ononis fruticosa.* France. Small shrub; 1 to 2 feet; rose; early summer; cuttings or seed.—Copses, open glades, or rough rocky ground.

Round-leaved Restharrow. *Ononis rotundifolia.* Switzerland. Small shrub; 1 to 2 feet; rose; summer; seed or cuttings.—Similar positions to the preceding.

Mountain Kidney Vetch. *Anthyllis montana.* Southern Europe. Herbaceous perennial; 2 to 4 inches; purplish; summer; seed, division, or cuttings.—Bare rocky ground and banks, or in short grass, in any soil.

Two-lobed Goat's Rue. *Galega biloba.* Persia. Herbaceous perennial; 2 to 3 feet; blue; summer; seed or division.—With the taller and handsomer-flowering herbs, in open spaces, in shrubberies, and on the margin of woods; also in long grass and on rough stony ground.

Officinal Goat's Rue. *Galega officinalis.* Southern Europe. Herbaceous perennial; 3 to 4 feet; lilac-purple; summer; seed or division.—Amidst long grass or vigorous herbs, in almost any position and soil, in copses, shrubberies, and open spots by wood walks.

Oriental Goat's Rue. *Galega orientalis.* Levant. Herbaceous perennial; 3 to 4 feet; blue; summer; seed or division.—Similar positions to those for the Officinal Goat's Rue and in ordinary soil.

Montpelier Milk Vetch. *Astragalus monspessulanus.* France. Evergreen perennial; 9 inches to 1 foot; red or purple; summer; seed or division.—Rocky places, banks

or slopes, where its prostrate shoots may show to greatest
advantage, in any soil.

Pontic Milk Vetch. *Astragalus ponticus.* Tauria.
Herbaceous perennial; 2 to 4 feet; pale yellow; summer;
seed or division.—Amidst vigorous perennials; chiefly
valuable for the effect of its handsome leaves on bold
stems; it thrives in any ordinary soil.

Rosy Coronilla. *Coronilla varia.* Europe. Herbaceous perennial; 1 to 2 feet; pink; summer; division.—
On grassy banks, stony heaps, rough rocky ground,
spreading over slopes or any like positions. A fine plant
for naturalization, thriving in any soil.

Canadian Desmodium. *Desmodium canadense.* North
America. Herbaceous perennial; 6 feet; purple; summer; seed or division.—Associated with the strongest
herbaceous plants in copses, woods, or any place where a
vigorous herbaceous vegetation is desired.

French Honeysuckle. *Hedysarum coronarium.*
Italy. Biennial; 3 to 5 feet; bright red; summer; seed.—
Somewhat open places in woods, shrubberies, and copses,
where it would sow itself.

Rock Hedysarum. *Hedysarum obscurum.* Europe.
Herbaceous perennial; 9 to 18 inches; purplish; sum-
mer; seed or division.—Positions similar to those for the
Rosy Coronilla; like it, a valuable plant for naturalization.

Silvery Vetch. *Vicia argentea.* Pyrenees. Herbaceous perennial; 9 to 15 inches; pink; summer; seed
or division.—Rocks, stony places, and thinly clad banks
in sandy, or ordinary soil.

Large Flowered Pea. *Lathyrus grandiflorus.*
Southern Europe. Climber; 3 to 4 feet; purple;

summer; seed or division.—Copses, fringes of woods, banks, hedges, margins of walks in the wilderness, or any position in which its free-flowering shoots may trail over shrubs, or fall over the face of rocks or banks: it will grow in almost any kind of soil.

Everlasting Pea. *Lathyrus latifolius.* Southern Europe. Climber; 4 to 8 feet; rose; all summer; seed or division.—Similar positions to the preceding. The white and the deeper-coloured varieties are even more beautiful than the common one.

Flaccid Bitter Vetch. *Orobus flaccidus.* Switzerland. Herbaceous perennial; 1 foot; blue and lilac; spring; seed or division.—In almost any soil or position where the vegetation does not exceed 18 inches.

Pea-like Bitter Vetch. *Orobus lathyroides.* Siberia. Herbaceous perennial; 1 to 2 feet; violet-blue; early summer; seed or division.—Similar positions to the preceding, but will thrive where the surrounding vegetation is taller, and in ordinary soil.

Spring Bitter Vetch. *Orobus vernus.* Europe. Herbaceous perennial; 1 to 2 feet; blue and lilac; spring; seed or division.—Banks, grassy unmown margins of wood walks, rocks, fringes of shrubberies, and like places; best in deep and sandy loam, well drained, though it will grow in almost any position.

Many-leaved Lupine. *Lupinus polyphyllus.* Columbia. Herbaceous perennial; 2 to 4 feet; blue; summer; seed or division.—Amidst the tallest and handsomest herbaceous plants, grouped where they may be seen from grass drives or wood walks, or in any position or soil. There are several varieties, all worthy of culture.

THE ROSE FAMILY.

Rosy Spiræa. *Spiræa venusta.* North America. Deciduous shrub; 2½ feet to 3½ feet; purplish-rose; summer; seed or division.—Associated with the handsomer and taller perennials.

Chili Avens. *Geum chilense.* Chili. Evergreen perennial; 2 to 3 feet; scarlet; summer; seed or division.—In any grassy or rocky places, or on banks.

Mountain Avens. *Geum montanum.* Alps and Pyrenees. Herbaceous perennial; 8 to 10 inches; yellow; summer; seed or division.—In upland pastures or boggy places, or on stony ground or banks.

Indian Strawberry. *Fragaria indica.* India. Trailing herb; 3 to 6 inches; yellow; summer; seed or runners.—Rocky places and banks, amidst dwarf Alpine plants and trailers in ordinary soil.

White-stemmed Bramble. *Rubus biflorus.* Western America. A vigorous erect bramble; white; summer; seed or cuttings.—Warm places in woods, copses, and on sunny wood banks, where its large white stems will show to great advantage; also fine for rocky places.

Nootka Sound Bramble. *Rubus nutkanus.* Northern America. Shrub; 3 to 6 feet; white; summer; seed or cuttings.—Almost any positions in woods, shrubberies; best near the walks; in all cases allowed to "run wild"—an excellent subject for naturalization.

Sweet-scented Bramble. *Rubus odoratus.* Northern America. Shrub; 4 to 6 feet; purplish-red; summer; seed or cuttings.—Similar positions to those for the preceding kind, and associated with them.

Showy Bramble. *Rubus spectabilis.* Northern America. Shrub; 4, 5, or 6 feet; dark purple; spring; seed or cuttings.—In warm sunny parts of sheltered shrubberies, where its early-blooming tendency may be encouraged; thrives freely in any soil.

Calabrian Cinquefoil. *Potentilla calabra.* Italy. Herbaceous perennial; 1 foot; yellow; summer; seed or division.—Rocky places and warm banks, where its prostrate silvery shoots may be seen to best advantage.

The numerous showy forms of Potentilla that may be easily raised from seed are all excellent for naturalization in any position or soil, and well able to take care of themselves among long grass and vigorous perennials.

THE EVENING PRIMROSE FAMILY.

Fraser's Evening Primrose. *Œnothera Fraseri.* Northern America. Herbaceous perennial; 1 to 2 feet; yellow; summer; seed or division.—Copses, grassy banks, and fringes of shrubberies in any soil.

James's Evening Primrose. *Œnothera Jamesi.* North America. Biennial; 4 feet; yellow; summer; seed.—Associated with vigorous herbaceous plants, in groups, near grass drives or wood walks. Very sweet in the evening, and worthy of being grown in quantity.

Missouri Evening Primrose. *Œnothera missouriensis.* North America. Herbaceous perennial; 1 foot; yellow; summer; seed or division.—Banks, edges of low shrubberies, fringes of copses, or rocky ground.

Lamarck's Evening Primrose. *Œnothera Lamarckiana.* North America. Biennial; 3 to 4 feet; yellow;

summer; seed.—A noble plant, suitable for the same positions as those recommended for Œ. Jamesi.

Swamp Evening Primrose. *Œnothera riparia.* North America. Herbaceous perennial; 1 to 2 feet; yellow; summer; seed or division.—Positions like those recommended for Œ. Fraseri.

Dandelion-leaved Evening Primrose. *Œnothera taraxacifolia.* Peru. Evergreen perennial; 6 to 9 inches; white; summer; seed or division.—The same positions as those recommended for the Missouri Evening Primrose.

Showy Evening Primrose. *Œnothera speciosa.* North America. Herbaceous perennial; 1 to 2 feet; white; summer; seed or division.—Rocky places, banks, and on the margins of shrubberies, in good light soil.

Lindley's Godetia. *Godetia Lindleyana.* North America. Annual; 2 feet; rosy-purple; summer; seed.—Banks, or any position in which vigorous annuals may be sown.

Reddish Godetia. *Godetia rubicunda.* California. Annual; 2 to 3 feet; red; summer; seed.—Similar position to the preceding, and in ordinary soil.

Elegant Clarkia. *Clarkia elegans.* California. Annual; 1 to 2½ feet; purple; summer; seed.—Should be sown with showy and vigorous annuals, like the preceding, on somewhat bare places in copses, and on slopes.

Pretty Clarkia. *Clarkia pulchella.* North America. Annual; 1 to 2½ feet; purple; summer; seed.—Similar positions to preceding, and like it grows in any soil.

THE PURSLANE FAMILY.

Umbelled Calandrinia. *Calandrinia umbellata.* Chili. Evergreen perennial; 3 to 6 inches; purple-crimson ; summer ; seed or division.—A gem for chinks in rocks, or on very sandy or peaty soils, in open and bare positions ; it perishes in winter on clay.

THE STONECROP FAMILY.

Ewers's Stonecrop. *Sedum Ewersii.* Siberia. Evergreen perennial ; 3 to 6 inches ; rose ; late summer ; seed or division.—Rocks and old walls.

Orange Stonecrop. *Sedum kamtschaticum.* Siberia. 3 to 6 inches ; summer ; deep orange ; seed or division. —Rocky and bare places, and banks.

Siebold's Stonecrop. *Sedum Sieboldii.* Japan. Perennial ; 2 to 4 inches ; pinkish ; late summer and autumn ; division.—Warm, rocky banks.

Great Stonecrop. *Sedum spectabile.* Japan. A stout perennial ; 15 to 24 inches ; rose ; autumn ; division or cuttings.—In any position where vigorous herbs may be grown ; best in open spots, associated with fine autumn plants, like Anemone japonica, the Tritomas, and the large Statice.

Webbed Houseleek. *Sempervivum arachnoideum.* Italy.—Evergreen perennial ; 3 to 6 inches ; red ; summer ; seed or division.—Rocks and bare stony banks, or on mossy old walls and ruins.

Glaucous Houseleek. *Sempervivum calcareum.* France. An evergreen plant with glaucous rosettes ; 2 to 8 inches ; pale rose : late in summer ; division.—In any position where the common Houseleek may be grown, and in any soil. A fine plant.

Hairy Houseleek. *Sempervivum hirtum.* Italy. Evergreen perennial ; 6 to 9 inches ; red ; summer ; seed or division.—In rocky or very bare places, in any aspect or soil. Bees are very fond of the flowers of this plant.

Mountain Houseleek. *Sempervivum montanum.* Switzerland. Evergreen perennial; 6 to 9 inches; red; summer; seed or division.—Same positions as for the preceding; both will thrive on walls.

Soboliferous Houseleek. *Sempervivum soboliferum.* Evergreen perennial; 4 to 6 inches; pale yellow; summer; seed or division.—Very bare and fully exposed places on rocky or stony ground. Should be allowed to spread into compact tufts.

THE SAXIFRAGE FAMILY.

Silvery Saxifrage. *Saxifraga Aizoon.* Pyrenees. Evergreen perennial; 6 to 9 inches; white spotted; early summer; seed or division.—Open bare spots on fully exposed rocky or bare ground, in any soil.

Two-flowered Saxifrage. *Saxifraga biflora.* Switzerland. Evergreen perennial; 2 to 4 inches; red; early summer; seed or division.—Moist rocky, bare spots, fully exposed, and in sandy peat or loam.

Heart-leaved Saxifrage. *Saxifraga cordifolia.* Siberia. Evergreen perennial; 9 to 18 inches; rose; spring and early summer; division.—Banks, rough rockwork, by wood walks, on wild, sunny slopes.

Thick-leaved Saxifrage. *Saxifraga crassifolia.* Siberia. Evergreen perennial; 1 to $1\frac{1}{2}$ feet; rose; spring and early summer; division.—Same situations as the preceding, and like it grows in any soil.

Crustate Saxifrage. *Saxifraga crustata.* Pyrenees. Evergreen perennial; 3 to 6 inches; whitish;

early summer; seed or division.—On walls, bare rocky
spots, or where there is a very dwarf and stunted vegetation; thrives best in a moist, sandy soil.

Juniper Saxifrage. *Saxifraga juniperina.* Caucasus. Evergreen perennial; 2 to 4 inches; yellow;
summer; division or seed.—Same positions as preceding,
but I have no proof that it would grow on walls.

Long - leaved Saxifrage. *Saxifraga longifolia.*
Pyrenees. Silvery perennial; 10 to 18 inches; white,
with pink spots; early summer; seed.—Walls, ruins,
rocks, and bare exposed spots, in ordinary soil.

Yellow Annual Saxifrage. *Saxifraga Cymbalaria.*
The East. Annual or biennial; 4 to 6 inches; yellow;
summer; seed.—Rocky, gravelly, or bare places, preferring a rather moist soil, and " sowing itself."

Pyramidal Saxifrage. *Saxifraga pyramidalis.* Europe. Evergreen perennial; 1 to 2 feet; white, with
reddish dots; summer; seed or division.—Same situations as those for the long-leaved Saxifrage.

Rivulet Astilbe. *Astilbe rivularis.* Nepal. Herbaceous perennial; 3 to 4 feet; whitish; summer;
division.—Among plants of striking habit or fine foliage,
by wood walks, or in glades—best in deep soil.

THE PARSLEY FAMILY.

Buenos Ayres Pennywort. *Hydrocotyle Bonariensis.*
Southern America. Trailer; 2 to 3 inches; green; summer;
seed or division.—Shrubberies, copses, or banks; desirable for the peculiarity of its leaves.

Greater Masterwort. *Astrantia major.* Europe.
Herbaceous perennial; 1 to 2 feet; striped red; sum-

mer ; seed or division.—Among the medium-sized herbaceous plants in glades, copses, and by wood walks.

Dwarf Dondia. *Dondia Epipactis.* Alps of Europe.
Herbaceous perennial ; 2 to 4 inches ; yellow; early
spring ; seed or division.—Banks, or anywhere amidst
a very dwarf vegetation ; grouped with early flowers.

Alpine Eryngo. *Eryngium alpinum.* Switzerland.
Herbaceous perennial ; 1 to 3 feet ; blue ; summer ; seed
or division.—In glades, copses, margins of shrubberies, etc.
A noble plant, thriving everywhere.

Amethystine Eryngo. *Eryngium amethystinum.*
Europe. Herbaceous perennial; 2 to 3 feet ; blue ;
summer ; seed or division.—Similar situations to preceding ; also worthy of extensive cultivation.

Tall Meadow Saxifrage. *Seseli elatum.* Austria.
Herbaceous perennial ; 1 to 2 feet ; white ; summer ;
seed or division.—Banks, wild walks, margins of shrubberies, etc. ; desirable for the beauty of its leaves.

Slender Meadow Saxifrage. *Seseli gracile.* Hungary.
Herbaceous perennial; 1 to 2 feet ; yellow ; summer ;
seed or division.—Similar uses to preceding.

Matthioli's Spignel. *Athamanta Matthioli.* Central
Europe. Herbaceous perennial ; 1 to 2 feet ; white ;
summer ; seed or division. Banks, rough rockwork, and
bare places. Valuable for its graceful tufts of leaves.

Cicuta-like Molopospermum. *Molopospermum cicu-
tarium.* Pyrenees and Alps. Herbaceous perennial ; 3
to 5 feet ; white ; early summer ; seed or division.—By
wood walks, among hardy plants with fine leaves or
striking habit, or isolated among flowering plants.

Common Giant Fennel. *Ferula communis.* Southern

Europe.   Herbaceous perennial; 8 to 12 feet; yellow; early summer; seed or division.—Isolated specimens by wood walks, and in glades, or grouped with other striking hardy plants. A noble plant.

Glaucous Giant Fennel. *Ferula glauca.* Southern Europe. Herbaceous perennial; 5 to 8 feet; pale yellow; early summer; seed or division.—Similar positions to the preceding; also a very remarkable plant.

Tangier Giant Fennel. *Ferula tingitana.* Southern Europe. Herbaceous perennial; 5 to 8 feet; yellow; summer; seed.—Another fine species, suitable for the same purposes, and thriving in ordinary soil.

Involucred Sulphurwort. *Peucedanum involucratum.* France. Herbaceous perennial; 3 to 6 feet; summer; seed or division.—Here and there among flowering plants for its graceful leaves and habit; on banks and bare glades in common, sandy soil.

Long-leaved Sulphurwort. *Peucedanum longifolium.* Hungary. Herbaceous perennial; 4 to 5 feet; yellow; summer; seed or division.—Similar uses to preceding. P. Petteri is also very suitable for like purposes; both thrive in common soil.

Giant Cow Parsnip. *Heracleum giganteum.* Siberia. Biennial; 6 to 10 feet; white; summer; seed.—Among the most vigorous herbaceous vegetation, in rich soil near river banks, or in any position where a striking distant effect is sought.

THE ARALIA FAMILY.

Naked-stalked Aralia. *Aralia nudicaulis.* North America. Herbaceous perennial; 4 to 5 feet; white;

summer ; division.—By wood walks, isolated, or grouped
with fine-foliaged herbaceous plants.

Berry-bearing Aralia. *Aralia racemosa.* North
America. Herbaceous perennial ; 4 to 5 feet ; white ;
summer ; division or seed.—Similar positions to preceding, in deep ordinary soil.

THE HONEYSUCKLE FAMILY.

Canadian Dogwood. *Cornus canadensis.* North
America. Herbaceous perennial ; 6 to 9 inches ; yellow ;
summer ; division.—Rocky and bare places, in sandy,
moist soil ; a singularly pretty plant.

Northern Linnæa. *Linnæa borealis.* Northern
Europe. Trailer ; 2 to 3 inches ; flesh-coloured ; all
summer ; division.—In moist rocky dells.

It need hardly be remarked that many of the shrubby
honeysuckles are among the most desirable subjects for
naturalization.

THE BEDSTRAW FAMILY.

Long-styled Crosswort. *Crucianella stylosa.* Persia.
Herbaceous perennial ; 1 to 1½ feet ; pink ; summer ;
seed or division.—Rocky places and banks, or in level
spots where the vegetation is dwarf.

THE VALERIAN FAMILY.

Red Valerian. *Centranthus ruber.* Europe. Herbaceous perennial ; 2 to 3 feet ; red ; all summer ; seed
or division.—This and its white variety are admirable for
banks, on which they frequently thrive far better than on
the level ground, though they thrive well almost anywhere.

THE TEASEL FAMILY.

Cut-leaved Teasel. *Dipsacus laciniatus.* Germany. Biennial; 4 to 6 feet; purple; summer; seed.—In open glades, and by wood walks, or on rich banks.

Long-leaved Morina. *Morina longifolia.* India. Evergreen perennial; 2 to 3½ feet; reddish; summer; seed or division.—Banks and margins of shrubberies, and on rough rockwork, near the eye.

Caucasian Scabious. *Scabiosa caucasica.* Caucasus. Herbaceous perennial; 1 to 2 feet; pale blue; summer; seed or division.—By wood walks, and on margins of shrubberies, in warm soil.

Grass-leaved Scabious. *Scabiosa graminifolia.* Switzerland. Herbaceous perennial; 1 foot; blue; summer; division or seed.—Rocky, or very bare places or banks; always in light warm soil.

THE COMPOSITE FAMILY.

Orange-coloured Hawkweed. *Hieracium aurantiacum.* Europe. Evergreen perennial; 1 to 1½ feet; orange; summer; seed or division.—In any position where the vegetation is not too coarse to hide it.

Plumier's Mulgedium. *Mulgedium Plumieri.* France. Herbaceous perennial; 4 to 6 feet; summer and early autumn; blue; seed or division.—Grouped with the most vigorous herbaceous plants, or as isolated tufts in wood walks, in deep rich soil.

Azure Catananche. *Catananche cærulea.* Southern Europe. Herbaceous perennial; 3 to 4 feet; blue; midsummer to autumn ; seed or division.— Banks, rocky ground, or on fringes of copses or woods.

Drooping Alfredia. *Alfredia cernua.* Siberia. Herbaceous perennial ; 3 to 5 feet ; yellow ; summer ; seed or division. Grouped with the stoutest and most vigorous herbaceous plants in wild places.

Illyrian Cotton Thistle. *Onopordum illyricum.* Southern Europe. Biennial ; 6 to 8 feet ; purplish ; summer ; seed. — Shrubberies, copses, or glades in woods.

French Artichoke. *Cynara Scolymus.* Southern Europe. Herbaceous perennial ; 4 to 8 feet ; purplish ; summer ; seed or division.—In glades near wood walks, or sloping ground, in dry deep soil. Most effective as isolated plants, though always very striking.

Babylonian Knapweed. *Centaurea babylonica.* Levant. Herbaceous perennial ; 5 to 10 feet ; yellow ; summer ; seed or division.—Associated with the most vigorous herbs, by wood walks, in any soil.

White-leaved Knapweed. *Centaurea dealbata.* Caucasus. Herbaceous perennial ; 2 to 3 feet ; purplish ; summer ; seed or division.—Fringes of shrubberies and on banks, in ordinary soil.

Mountain Knapweed. *Centaurea montana.* Austria. Herbaceous perennial ; 2 to 3 feet ; blue ; summer ; seed or division.—Similar positions to the preceding.

One-flowered Knapweed. *Centaurea uniflora.* Southern Europe. Herbaceous perennial ; 1 to 2 feet ; purple ; summer and autumn ; seed or division.—Rocky places and banks. A handsome mountain plant.

Hungarian Globe Thistle. *Echincps bannaticus*

G

Hungary. Herbaceous perennial; 5 to 8 feet; blue; summer; seed or division.—Woods, copses, or by pleasure-ground walks; and also for association with herbaceous plants of some vigour and character.

Tall Globe Thistle. *Echinops exaltatus.* Austria. Herbaceous perennial; 6 to 8 feet; white; summer; seed or division.—Similar uses to the preceding, and suited for association with even more vigorous vegetation.

Russian Globe Thistle. *Echinops ruthenicus.* Russia. Herbaceous perennial; 6 to 8 feet; blue; summer; seed or division.—A fine subject for planting in tufts in open spots by wood walks, and for association with the handsomest and most vigorous herbaceous flowering plants.

Elegant Liatris. *Liatris elegans.* North America. Herbaceous perennial; 3 to 4 feet; purple; autumn; seed or division.—Fringes of shrubberies, in open sunny places on warm soils.

Dotted Liatris. *Liatris punctata.* North America. Herbaceous perennial; 4 feet; reddish-purple; summer; seed or division.—Similar uses to the preceding.

Long-spiked Liatris. *Liatris spicata.* North America. Herbaceous perennial; 4 feet; late summer and early autumn; seed or division.—Similar uses to the preceding.

Caucasian Leopard's Bane. *Doronicum caucasicum.* Caucasus. Herbaceous perennial; 1 to 2 feet; yellow; spring and summer; seed or division.—Rocky places, banks, fringes of low shrubberies in any soil. Best in sunny positions, as it flowers early in the year.

Pearl Cudweed. *Antennaria margaritacea.* North America. Herbaceous perennial; 1 to 2 feet; white; summer; seed or division.—Banks, open spots in copses or fringes of low shrubberies, in any soil.

Lion's Foot Everlasting. *Gnaphalium Leontopodium.* Switzerland. Herbaceous perennial; ½ foot; white, summer; seed or division.—Bare rocky places, in moist sandy soil, amidst vegetation not over 6 inches high. A most interesting plant for naturalization, in upland districts, especially where the rock crops out.

Yellow Everlasting. *Helichrysum arenarium.* Europe. Herbaceous perennial; 6 to 12 inches; yellow; summer; seed or division.—Bare rocky places or banks, on a sandy warm soil, always in positions where it may not be overrun by other plants. Perishes on cold clay soils.

Showy Stenactis. *Stenactis speciosa.* California. Herbaceous perennial; 2 feet; purplish; summer; seed or division.—Among medium-sized, choice herbaceous plants, on the fringes of shrubberies, banks, or in rocky places, in ordinary soil.

Alpine Starwort. *Aster alpinus.* Europe. Herbaceous perennial; 6 to 12 inches; purplish; summer; seed or division.—An interesting plant for naturalization in upland meadows, or in the rougher grassy parts of pleasure grounds, as it grows abundantly in many sub-alpine pastures, always much smaller than when grown in rich garden soil; will also suit rocky places, banks, or borders, among plants not more than a foot high.

Italian Starwort. *Aster Amellus.* Southern Italy. Herbaceous perennial; 2 feet; purple; summer; seed or division.—Banks, fringes of shrubberies, or rough rocky places, in almost any soil.

Heart-leaved Starwort. *Aster cordifolius.* North America. Herbaceous perennial; 3 to 4 feet; purplish; summer; seed or division.—Open spaces, glades in woods, by wood walks, and on banks, in any soil.

Spreading Starwort. *Aster diffusus.* North America. Herbaceous perennial; 3 to 5 feet; white; autumn; seed or division.—Similar uses to the preceding.

Heath-like Starwort. *Aster ericoides.* Herbaceous perennial; 3 to 5 feet; white; late summer; seed or division.—Banks, rough rocky places, fringes of woods or copses, in ordinary soil.

Many-flowered Starwort. *Aster floribundus.* North America. Herbaceous perennial; 4 to 5 feet; purple; late summer and autumn; seed or division.—Similar positions to preceding.

New York Starwort. *Aster Novi Belgii.* North America. Herbaceous perennial; 4 to 6 feet; pale blue; autumn; division.—Association with the most vigorous herbaceous plants, in copses, fringes of woods, or in glades, flowering best in sunny spots.

New England Starwort. *Aster Novæ Angliæ.* North America. Herbaceous perennial; 6 feet; purple; autumn; division.—Similar positions to preceding, and suitable for association with it.

Pyrenean Starwort. *Aster pyrenæus.* Pyrenees. Herbaceous perennial; 4 to 6 feet; violet; summer; division.—A very handsome summer-flowering kind, suited for banks, fringes of shrubberies, and rocky ground, in any position, cold or hot.

Australian Daisy. *Vittadenia triloba.* Australia. Evergreen perennial; 1 foot; pale lilac; all summer;

seed.—This pretty plant may be naturalized on the warm slopes of old quarries, &c., in the southern and milder parts of the country, in light or stony soil.

Golden Rod. *Solidago.* The numerous species of this genus, yellow-flowered, and for the most part tall and vigorous herbs, are well fitted for naturalization in woody places and copses. Indeed, wild or semi-wild places are the only ones for which they are suited. They are all as easily propagated as the Michaelmas Daisy, and will grow in any soil.

Ageratum Eupatory. *Eupatorium ageratoides.* North America. Herbaceous perennial ; 4 feet ; white ; late in summer ; seed or division.—Similar positions to those for the medium-sized Michaelmas Daisies, and suited for association with them in any soil.

Purple-stalked Eupatory. *Eupatorium purpureum.* North America. Herbaceous perennial ; 4 to 5 feet ; purple ; autumn ; seed or division.—Fringes of woods and shrubberies, in sunny aspects.

Winter Heliotrope. *Tussilago fragrans.* Europe. Herbaceous perennial ; 1 foot ; blush ; winter ; division.— Naturalization in any wild places in the shade of trees, on banks, in lanes, and neglected places. Valuable for cutting from in winter or early in spring, but being a fast-spreading "weed" should not be planted where its rapid increase could prove objectionable.

Sea Ragwort. *Cineraria maritima.* Southern Europe. Evergreen perennial ; 2 feet ; yellow ; late summer ; seed.—Dry sandy banks, in warm rocky spots, old quarries, &c. In some parts it will survive on level ground, especially in very light soils.

Glaucous Ragwort. *Othonna cheirifolia.* Barbary. Evergreen perennial; 1 foot; yellow; early summer; cuttings.—Banks, borders, or rocky places. Sometimes perishes in winter on the level ground in very cold soils, but generally free and hardy.

Heart-leaved Telekia. *Telekia cordifolia.* Hungary. Herbaceous perennial; 3 to 5 feet; yellow; summer; division.—A vigorous herbaceous plant, suited for association with Echinops, Rheum, and subjects grown for their foliage and character.

Double Sunflower. *Helianthus multiflorus.* North America. Herbaceous perennial; 6 to 8 feet; yellow; summer; seed or division.—The double variety of this is that most commonly seen. It is very ornamental, and will thrive in almost any soil, in woods and copses.

Drooping-leaved Sunflower. *Helianthus orgyalis.* North America. Herbaceous perennial; 9 to 10 feet; orange-yellow; autumn; seed or division.—An exceedingly graceful plant, when seen in an isolated tuft, near a wood walk. Also suited for association with plants of fine leaf and character, in rich soil.

Newman's Rudbeckia. *Rudbeckia Newmani.* South America. Herbaceous perennial; 1 to 2 feet; yellow; summer; seed or division.—A very showy vigorous plant; fine for fringes of shrubbery, copses, and groups of late perennials.

Rigid Sunflower. *Harpalium rigidum.* North America. Herbaceous perennial; $3\frac{1}{4}$ feet; dark yellow; late summer; division.—A brilliant and showy free-running perennial, excellent for copses, fringes of woods, or almost any position in the wilder parts of a country place, thriving in ordinary soil.

Common Marigold. *Calendula officinalis.* Southern Europe. Annual; 1½ feet; orange; all summer; seed. —Suited for almost any position or soil in semi-wild places, not overrun by very coarse vegetation.

Cup Plant. *Silphium perfoliatum.* North America. Herbaceous perennial; 5 to 7 feet; yellow; summer; seed or division.—Association with the tallest and most vigorous herbs in rich or deep soil.

Alpine Lavender Cotton. *Santolina alpina.* Southern Europe. Evergreen perennial; 3 to 6 inches; yellow; late in summer; division or cuttings.—Very bare rocky places, or banks, amidst dwarf rock plants.

Ground Cypress. *Santolina Chamæcyparissus.* Southern Europe. Evergreen perennial; 2 to 3 feet; yellow; summer; cuttings.—A fine plant for banks, rocky places, or the outer fringes of shrubberies.

Hoary Lavender Cotton. *Santolina incana.* Southern Europe. Evergreen perennial; 1 foot; yellow; summer; seed or cuttings.—A variety of the preceding, but dwarfer; suited for rather bare banks and rocky places, in ordinary soil.

Hoary Wormwood. *Artemisia cana.* North America. Shrub; 1 to 3 feet; yellowish; summer; division or seed. — Fringes of woods and shrubberies, or rough rocky ground or banks. A vigorous silvery-leaved plant.

Cape Marigold. *Dimorphotheca pluvialis.* Cape of Good Hope. Annual; 1 to 2 feet; whitish-purple;

summer; seed.—Sunny and rather bare banks, or slopes, in somewhat dry and good soil.

Noble Achillea. *Achillea Eupatorium.* Shores of the Caspian. Evergreen perennial; 3 to 5 feet; yellow; summer; seed or division.—Fringes of shrubberies, and associated with the noblest herbaceous plants, in almost any position or soil.

Rosy Feverfew. *Pyrethrum roseum.* Caucasus. Evergreen perennial; 2 feet; rose; summer; seed or division.—The numerous single and double varieties of this fine hardy plant grow freely in almost any soil or position, but are most suitable for the low fringes of shrubberies, or low-lying banks, in rich soil.

Marsh Feverfew. *Pyrethrum uliginosum.* Hungary. Herbaceous perennial; 3 to 4 feet; white; late in summer; seed or division.—A showy late-flowering plant, fine for grouping with the best Michaelmas daisies and other large and effective plants, in rich moist soil.

THE BELL-FLOWER FAMILY.

Pendulous Bell-flower. *Symphyandra pendula.* Caucasus. Herbaceous perennial; 1 to 1½ feet; white; summer; seed or division.—Banks or rocky places, amidst vegetation not more than about 1 foot high.

Bearded Bell-flower. *Campanula barbata.* Italy. Herbaceous perennial; 1 foot; pale blue; summer; seed or division.—Banks, rocky places, or in grass.

Carpathian Bell-flower. *Campanula carpatica.* Mid-Europe. Herbaceous perennial; 1 foot; blue; summer;

seed or division.—A lovely plant for banks, rocky places, or any position in which it may not be overrun by taller vegetation. It thrives in any soil.

Fragile Bell-flower. *Campanula fragilis.* Italy. Herbaceous perennial; ½ foot; pale blue; summer; seed or division.—On dry sunny banks, amidst dwarf rock plants, or in crevices in old quarries, &c.

Gargano Bell-flower. *Campanula garganica.* Italy. Herbaceous perennial; 3 to 6 inches; blue; summer; seed or division.—Similar uses to the preceding.

Tall Bell-flower. *Campanula grandis.* Asia Minor. Herbaceous perennial; 3 feet; blue; summer; seed or division.—Association with the finer medium-sized herbaceous plants, on fringes of shrubberies, and in open glades in woods.

Equal-leaved Bell-flower. *Campanula isophylla.* North Italy. Herbaceous perennial; 3 to 6 inches; blue; summer; seed or division.—Similar positions to those for C. fragilis, or C. garganica.

Wall Bell-flower. *Campanula muralis.* Dalmatia. Herbaceous perennial; 8 to 12 inches; pale blue; summer; seed or division.—Rocky places or old quarries; if possible, in chinks against a vertical face of rock, where it will prove most ornamental.

Long Bell-flower. *Campanula nobilis.* China. Herbaceous perennial; 1 to 2 feet; purple or white; summer; seed or division.—Banks and rocky places, or on the level ground, on which, however, its large pendulous blossoms will not be seen to such great advantage as when the plant is somewhat elevated.

Peach-leaved Bell-flower. *Campanula persicifolia.* Southern Europe. Herbaceous perennial; 2½ feet; blue; summer; division.—Similar positions to those for C. grandis.

THE HEATH FAMILY.

Spring Heath. *Erica carnea.* Germany. Small shrub; 6 to 12 inches; pale purple; winter; division or cuttings.—Among our wild heaths, or on margins of shrubberies, in ordinary garden soil, though it thrives best in peat.

Empetrum-like Menziesia. *Menziesia empetriformis.* North America. Small shrub; pale red; summer; division or cuttings.—Very bare rocky places, in moist peaty soil; chiefly suited for moist or upland districts, unless when carefully grown in the rock-garden.

Partridge Berry. *Gaultheria procumbens.* North America. Trailing shrub; 2 to 4 inches; white; summer; division.—Rocky and bare places, or in almost any position amidst very dwarf vegetation; always in rather moist soil; best in that which is somewhat peaty.

THE PERIWINKLE FAMILY.

Herbaceous Periwinkle. *Vinca herbacea.* Hungary. Perennial; 1 to 1½ feet; blue; summer; division.—Woods, copses, fringes of shrubberies, banks, rough rockwork, &c.

Greater Periwinkle. *Vinca major.* Southern Europe. Trailer; 1 to 1½ feet; blue; summer; division.— Similar positions to preceding, but this is a much stronger plant. The fine variegated kinds thrive perfectly in woods and semi-wild spots, in any kind of soil.

Lesser Periwinkle. *Vinca minor.* Europe. Trailer ; 1 foot ; blue ; summer ; division.—Similar positions to preceding. There are many varieties, all good.

THE SWALLOWWORT FAMILY.

Douglas's Swallowwort. *Asclepias Douglasii.* North America. Herbaceous perennial ; 4 feet ; purple ; summer ; seed or division.—Among the finer herbaceous plants in copses, on margins of shrubberies, or on banks, in rich and deep soil.

Cornuti's Swallowwort. *Asclepias Cornuti.* North America. Herbaceous perennial ; 5 feet ; purple ; summer ; seed or division.—Similar positions to preceding ; but this plant is suited for association with the tallest and most vigorous herbaceous subjects.

THE GENTIAN FAMILY.

Stemless Gentian. *Gentiana acaulis.* Central Europe. Evergreen perennial ; 2 to 4 inches ; blue ; early summer ; division.—Bare rocky places ; seldom thrives in dry soil. Would grow on bare ground on our mountains or high hills as well as it does on the Alps.

Swallowwort Gentian. *Gentiana asclepiadea.* Austria. Herbaceous perennial ; 1 to 2 feet ; blue ; summer ; seed or division.—Amidst the finer dwarf herbaceous plants, on the margins of woods and copses, on rough rockwork, &c., in ordinary soil ; best, however, in peat or very light sandy loam.

Worm Grass. *Spigelia marilandica.* North America. Herbaceous perennial; 1 to 1½ feet; red; summer; division.—Now perhaps too rare for naturalization but well worthy of trial where there is any moist, sandy, peat soil, in a semi-wild place, on the fringes of copses, or open bare glades in woods.

THE PHLOX FAMILY.

Creeping Phlox. *Phlox reptans.* North America. Evergreen perennial; 3 to 6 inches; reddish; spring; division.—Bare rocky places, fringes of low copses, or wherever there is a very dwarf vegetation.

Awl-leaved Phlox. *Phlox subulata.* North America. Evergreen perennial; 4 inches; pink; early summer; division.—Similar positions to the preceding.

THE BINDWEED FAMILY.

Lined Bindweed. *Convolvulus lineatus.* Southern Europe. Trailer; 3 to 6 inches; bluish; summer; division.—In sandy soil, amidst the very dwarfest vegetation, on slopes, banks, &c.

Dahurian Bearbind. *Calystegia dahurica.* Dahuria. Climber; 1 to 5 feet; pink; summer; division.—In copses, hedges, over old stumps, railings, &c. A lovely twining plant, hardy and vigorous.

Pubescent Bearbind. *Calystegia pubescens.* China. Climber; 10 to 15 feet; pale rose; summer; division.—Similar uses to preceding; plant not so vigorous.

THE BORAGE FAMILY.

Prostrate Gromwell. *Lithospermum prostratum.*
Southern Europe. Evergreen trailer; 6 to 12 inches;
blue; summer; cuttings.—Rocky or bare places, margins
of copses, banks, &c.; flourishes best in deep, well-
drained, sandy loam.

Rough Comfrey. *Symphytum asperrimum.* Caucasus. Herbaceous perennial; 4 to 6 feet; blue;
summer; seed or division. — In woods and rough
shrubberies. A grand plant for naturalization.

Bohemian Comfrey. *Symphytum bohemicum.* Bo-
hemia. Herbaceous perennial; 3 feet; red; early
summer; division.—Copses, margins of plantations, by
wood walks, &c. in ordinary soil.

Oriental Comfrey. *Symphytum orientale.* Tauria.
Herbaceous perennial; 2 to 3 feet; white; early summer;
seed or division.—Similar positions to those for the Bohemian Comfrey.

Caucasian Comfrey. *Symphytum caucasicum.* Caucasus. Herbaceous perennial; 3 to 4 feet; blue; summer; division.—In shrubberies, copses, fringes of woods,
&c., also in more open positions amongst medium-sized
herbs. A lovely plant for naturalization, and thriving
freely in any soil.

Italian Bugloss. *Anchusa italica.* Southern Europe.
Herbaceous perennial; 2 to 3 feet; blue; summer;
seed or division.—Fringes of woods, copses, &c., or
among the stronger herbaceous plants.

Azorean Forget-me-not. *Myosotis azorica.* Azores.

Biennial; 1 foot; dark blue; autumn; seed.—In warm nooks in sandy, moist soil.

Early Forget-me-not. *Myosotis dissitiflora.* Alps. Herbaceous perennial; 6 to 12 inches; blue; early spring; seed or division.—Rocky places, banks, fringes of shrubberies and thin places in copses.

Creeping Forget-me-not. *Omphalodes verna.* Alps of Europe. Evergreen perennial; 6 inches; blue; spring; division.—In rocky places, fringes of low shrubberies, open spots in copses, by wood walks, &c.; prefers a somewhat moist soil.

Apennine Hound's Tongue. *Cynoglossum apenninum.* Italy. Herbaceous perennial; 1 to 2 feet; blue; summer; division or seed.—Amongst herbs from 1 foot to 18 inches high, on margins of shrubberies.

Cretan Borage. *Borago cretica.* Crete. Herbaceous perennial; 1 to 2 feet; blue; spring; division.—By wood walks, or in lawns in quiet shady places.

Oriental Borage. *Borago orientalis.* Turkey. Herbaceous perennial; 2 to 3 feet; blue; summer; division.—Similar positions to preceding.

Loose-flowered Borage. *Borago laxiflora.* Corsica Biennial; 6 to 12 inches; blue; summer; seed.—In bare places, banks, &c.; best in sandy soil; too small for association with the other Borages and Comfreys recommended; sows itself freely.

THE NIGHTSHADE FAMILY.

Pyrenean Ramondia. *Ramondia pyrenaica.* Pyrenees. Herbaceous perennial; 4 inches; purple; sum-

mer ; division.—Moist and warm rocky spots facing
south, in woods or shrubberies ; should be isolated from
coarse or creeping plants, and put in spongy loam or
peat, and never among coarse plants.

Winter Cherry. *Physalis Alkekengi.* Southern Europe. Herbaceous perennial ; 1 foot ; white ; late summer ; seed or division.—Fringes of copses, or banks near
wood walks ; best in warm soil.

THE FIGWORT FAMILY.

Perennial Mullein. *Verbascum Chaixii.* Southern
France. Herbaceous perennial ; 4 to 6 feet ; yellow, with
brown and purple centre ; summer ; seed or division.—
A few feet or yards within fringes of woods, or associated
with the largest and handsomest herbaceous plants.
Other large kinds are good, but the preceding is a true
perennial.

Great Snapdragon. *Antirrhinum majus.* Europe.
Evergreen perennial ; 2 feet ; red ; summer ; seed or
cuttings.—Rocky places. Although this grows in almost
any soil, it is on walls and ruins that it becomes
thoroughly established.

Rock Snapdragon. *Antirrhinum rupestre.*—Perennial ; ½ foot ; purplish-pink ; summer ; seed.—Rocky and
bare places, walls, and on banks.

Alpine Toadflax. *Linaria alpina.* Austria. Evergreen perennial ; 6 to 12 inches ; violet ; summer ; seed.
In bare, open, sandy, gritty, or gravelly spots, in the
moister and more elevated districts.

Broom-leaved Toadflax. *Linaria genistæfolia.* Austria. Herbaceous perennial; 2 feet; yellow; summer; seed or division.—Banks or copses.

Hartweg's Pentstemon. *Pentstemon Hartwegi.* Mexico. Evergreen perennial; 2 feet; red; summer; cuttings or seed.—Margins of shrubberies, open places in copses and on banks; growing best and enduring longest in a light rich soil, not very wet and cold in winter.

Tufted Pentstemon. *Pentstemon procerus.* North America. Evergreen perennial; 1 foot; purple; summer; seed or division.—Bare and rocky places, or almost anywhere amidst very dwarf vegetation, in ordinary soil.

Bearded Chelone. *Chelone barbata.* Mexico. Herbaceous perennial; 2 to 3 feet; orange-scarlet; summer; seed or division.—Among the finer herbaceous plants on margins of shrubberies, in copses, or in rocky places. The variety Torreyi is very large and fine.

Common Musk. *Mimulus moschatus.* Columbia. Perennial; 9 to 12 inches; yellow; summer; seed or division.—Best in somewhat shady positions, but will grow anywhere.

Many of the varieties of Monkey flower (*Mimulus*) are more ornamental than the species in cultivation. They may be naturalized in moist rich soil, on margins of shrubberies, and beds of American plants, or in open shrubby or heathy places, where there is a moist soil.

Alpine Erinus. *Erinus alpinus.* Pyrenees. Evergreen perennial; 3 to 4 inches; rosy purple; summer; seed.—Walls and ruins : grows better in these positions than on the level ground.

Amethystine Speedwell. *Veronica amethystina.*

Southern Europe. Herbaceous perennial; 3 feet; blue; summer; seed or division.—Margins of shrubberies, copses, or anywhere associated with the medium-sized herbaceous plants in any soil.

Austrian Speedwell. *Veronica austriaca.* Austria. Herbaceous perennial; 1 foot; light blue; summer; seed or division.—With the neater herbaceous plants on margins of shrubberies, banks, and slopes.

Hoary Speedwell. *Veronica incana.* Russia. Herbaceous perennial; 2 feet; blue; summer; seed or division.—Rocky places, or bare banks in ordinary soil.

THE SAGE FAMILY.

Lemon Thyme. *Thymus citriodorus.* Dwarf evergreen; 4 to 6 inches; flowers inconspicuous; division.—Rocky and bare places, dry, sandy, or gravelly banks.

Corsican Thyme. *Thymus corsicus.* Corsica. Evergreen perennial; 1 inch; lilac; summer; seed or division.—A very diminutive, strongly peppermint-scented plant, that will creep about amidst the very dwarfest vegetation in rocky places, and on moist banks.

Variegated Common Garden Thyme. *Thymus vulgaris.* Variegated garden variety; evergreen perennial; 1 foot; purplish; summer; division.—Banks or rocky places, in dry soil.

Common Germander. *Teucrium Chamædrys.* Europe. Herbaceous perennial; ¾ foot; purple; summer; division.—Banks and fringes of shrubberies, among the dwarfer herbaceous plants.

Hyrcanian Germander. *Teucrium hyrcanicum.*

Persia. Herbaceous perennial; 1 to 3 feet; purple; summer; division or seed.—Copses, bare openings in shrubberies, among the stronger herbaceous plants.

Geneva Bugle. *Ajuga genevensis.* Switzerland. Herbaceous perennial; ½ foot; flesh; summer; seed or division.—Margins of shrubberies, banks, or anywhere amongst vegetation not above 1 foot high, in ordinary soil.

Oswego Tea. *Monarda didyma.* North America. Herbaceous perennial; 2 to 3 feet; red; summer; seed or division.—Margins of shrubberies, copses, or open spots in glades, associated with large herbaceous plants; thrives best in sandy loam.

Hollow-stemmed Monarda. *Monarda fistulosa.* North America. Herbaceous perennial; 2 to 4 feet; purplish; summer; seed or division. There are several varieties.—Similar soil and positions to preceding.

Kalm's Monarda. *Monarda Kalmiana.* North America. Herbaceous perennial; 3 to 4 feet; purple; summer; division or seed.—Similar soil and positions to those recommended for the two preceding kinds.

Wind-herb Phlomis. *Phlomis herba-venti.* Southern Europe. Evergreen perennial; 1 to 2 feet; red; late summer; seed or division.—Sloping banks, margins of shrubberies, in tufts by wood walks, in ordinary soil.

Russell's Phlomis. *Phlomis Russelliana.* Levant. Herbaceous perennial; 3 to 4 feet; brownish; early summer; seed or division.—Similar positions to the preceding, but may be put among bolder vegetation.

Spotted Archangel. *Lamium maculatum.* Italy. Herbaceous perennial; 1 foot; purple; summer; division.—Fringes of plantations, banks, and rocky places. The white variety is very showy and good.

Showy Physostegia. *Physostegia speciosa.* North America. Herbaceous perennial; 2 to 3 feet; pink; summer; seed or division.—Among the finer perennials on the margins of shrubberies, in copses, or in groups by wood walks.

Imbricated Physostegia. *Physostegia imbricata.* North America. Herbaceous perennial; 4 to 6 feet; rose; summer; seed or division.—Similar positions to preceding, but suited also for association with larger plants.

Virginian Physostegia. *Physostegia virginiana.* North America. Herbaceous perennial; 2 to 3 feet; red; summer; seed or division.—Similar positions to those for P. speciosa.

Woolly Woundwort. *Stachys lanata.* Siberia. Herbaceous perennial; 1 to 2 feet; purple; summer; division or seed.—Margins of shrubberies or on banks in any kind of soil.

Lavender-leaved Zietenia. *Zietenia lavandulæfolia.* Levant. Evergreen perennial; 1 foot; purple; summer; division.—Similar positions to preceding; very free on warm and sandy soils.

Fisher's Dragon's Head. *Dracocephalum argunense.* Siberia. Herbaceous perennial; 1 foot; blue; summer; division or seed.—Rocky places and bare banks, in sandy soil or free loam.

Austrian Dragon's Head.   *Dracocephalum austri-acum.* Austria. Herbaceous perennial; 1 to 2 feet; blue; summer; seed or division.—Similar soil and positions to preceding.

Prickly-leaved Dragon's Head.   *Dracocephalum peregrinum.* Siberia. Herbaceous perennial; 1 foot; blue; summer; seed or division.—Banks and rocky places, amongst dwarf vegetation, in light, well-drained soil.

Common Balm.   *Melissa officinalis.*   Southern Europe. Herbaceous perennial; 1 foot; white; all summer; seed or division.—Margins of shrubberies, copses, or in tufts by wood walks. Only desirable for the odour of its leaves.

Alpine Skullcap.   *Scutellaria alpina.* Hungary. Herbaceous perennial; 1 to 2 feet; blue and white; summer; seed or division.—Dry sandy slopes or banks, on margins of shrubberies. Best in warm sandy loam, but easy to grow in ordinary soil.

Tartarian Skullcap.   *Scutellaria lupulina.* Tartary. Herbaceous perennial; 1 to 2 feet; yellow; summer; seed or division.—Similar soil and positions to preceding.

Silvery Sage.   *Salvia argentea.* Crete. Herbaceous perennial; 3 feet; white; early summer; seed.—Rocky places or banks, in good soil.

Wild Sage.   *Salvia sylvestris.* Europe. Herbaceous perennial; 3 feet; purplish; summer; seed or division.— In copses, &c., and on rough banks, amongst the coarsest herbaceous plants.

Large-flowered Self-heal. *Prunella grandiflora.* Europe. Herbaceous perennial; ¾ to 1½ feet; blue; summer; seed or division.—Fringes of shrubberies, banks, rocky places, &c., in soils not too wet in winter.

THE VERVAIN FAMILY.

Knot-flowered Zapania. *Zapania nodiflora.* South America. Evergreen perennial; 3 to 6 inches; pink; summer; division or cuttings.—Banks and rocky places, amidst dwarf trailing herbs.

THE BEAR'S BREECH FAMILY.

Broad-leaved Bear's Breech. *Acanthus latifolius.* Portugal. Herbaceous perennial; 2 to 4 feet; purplish; summer; seed or division.—In isolated tufts by wood walks, or grouped with herbaceous plants having fine foliage, always in deep rich soil.

Soft Bear's Breech. *Acanthus mollis.* Italy. Herbaceous perennial; 3 feet; pale purple; summer; seed or division.—Similar soil and positions to preceding.

Bristling Bear's Breech. *Acanthus spinosissimus.* Southern Europe. Herbaceous perennial; 3 feet; pale purple; summer; division or seed.—Similar soil and positions to preceding. Not so vigorous, but a very singular and desirable plant.

Spiny Bear's Breech. *Acanthus spinosus.* Southern Europe. Herbaceous perennial; 3 to 4 feet; pale purple; summer; seed or division.—Similar soil and positions to preceding.

THE PRIMROSE FAMILY.

European Cyclamen. *Cyclamen europæum.* Europe. Tuber; 4 inches; light red; summer; seed.—Rocky places and banks amidst very dwarf plants.

Ivy-leaved Cyclamen. *Cyclamen hederæfolium.* Europe. Tuber; 4 inches; purplish; summer; seed.— Similar positions to preceding. I have seen it naturalized with success in woods under the shade of high trees. In moss or short grass, but not in the neighbourhood of coarse herbs, brambles, &c. Other kinds of Cyclamen may be tried, notably C. coum and C. vernum.

Jeffrey's American Cowslip. *Dodecatheon Jeffreyi.* North America. Herbaceous perennial; 1½ feet high; purplish; early summer; seed or division.—Rocky places or low banks, in rich light and deep soil. A noble plant, as yet rare.

Mead's American Cowslip. *Dodecatheon Meadia.* North America. Herbaceous perennial; 1 to 1½ feet; purplish; early summer; seed or division.—Similar soil and positions to those for D. Jeffreyi. Not so vigorous.

Small American Cowslip. *Dodecatheon integrifolium.* North America. Herbaceous perennial; ½ foot; pale purple; spring; seed or division.—Rocky places or banks, in moist sandy or peaty soil and amid dwarf plants.

Alpine Soldanella. *Soldanella alpina.* Switzerland. Evergreen perennial; ¼ foot; purplish blue; summer; seed or division.—Bare rocky places, amidst minute vegetation, in moist very sandy soil.

Auricula. *Primula Auricula.* Switzerland. Evergreen perennial; ½ foot; various; spring; seed or division.—Rocky or bare places. Might be naturalized in upland districts wherever the grass is rather short.

Snowy Primrose. *Primula nivalis.* Dahuria. Evergreen perennial; ½ foot; white; early summer; division or seed.— Bare rocky places, in humid, elevated parts of the country, in moist sandy or peaty soil.

Fairy Primrose. *Primula minima.* Southern Europe. Evergreen perennial; 2 to 3 inches; purple; spring; seed or division.—Very bare rocky places, in northern and elevated parts of the kingdom only, associated with such dwarf plants as P. scotica, and the Pinguiculas.

Large-leaved Primrose. *Primula Palinuri.* Evergreen perennial; 1 foot; yellow; early summer; division.—Similar soil and positions to those recommended for the American cowslips.

Sikkim Primrose. *Primula sikkimensis.* Himalayas. Evergreen perennial; 1 to 2 feet; yellow; summer; division or seed.—At present a somewhat scarce plant. May, when sufficiently plentiful, be naturalized in rocky places, or on low banks in somewhat sheltered positions in moist, deep and light soils.

Long-flowered Primrose. *Primula longifolia.* Europe. Herbaceous perennial; 3 to 6 inches; red; summer; division or seed.—May be naturalized in positions, and under conditions, in which our own Bird's-eye Primula is found to thrive.

Clammy Primrose. *Primula viscosa.* Piedmont. Evergreen perennial; 3 to 6 inches; purple; spring;

division or seed.—Similar soil and positions to those re-commended for the Snowy Primrose.

Rock jasmine. *Androsace Chamæjasme.* Austria. Evergreen perennial; 2 to 3 inches; pink; summer; seed or division.—Similar positions and soil to those recommended for the Fairy Primrose.

Woolly-leaved Androsace. *Androsace lanuginosa.* Himalayas. Evergreen perennial; ½ foot; lilac; summer; division or seed.—Rocky bare places and banks only, in the southern and milder parts of the country, and amongst very dwarf trailing herbs, in free soil.

Yellow Androsace. *Androsace Vitaliana.* Pyrenees. Evergreen perennial; 3 inches; yellow; early summer; division.—Positions and localities similar to those re-commended for the Fairy Primrose, in light and moist soil.

Willow-leaved Loose-Strife. *Lysimachia Epheme-rum.* Spain. Herbaceous perennial; 2 to 3 feet; white; summer; division.—Margins of plantations and copses.

THE PLUMBAGO FAMILY.

Broad-leaved Sea Lavender. *Statice latifolia.* Siberia. Evergreen perennial; 2 to 3 feet; blue; sum-mer; seed or division.—Isolated tufts in glades by wood walks, and also associated with the finer autumnal flowering herbs in almost any position.

Tartarian Sea Lavender. *Statice tatarica.* Russia. Evergreen perennial; 2 to 3 feet; pink; summer; seed

or division.—Rocky places amidst vegetation not much
over a foot high.

Prickly Thrift. *Acantholimon glumaceum.* Armenia.
Evergreen perennial; 6 to 9 inches; rose; summer; seed
or cuttings.—Bare rocky places, banks or slopes, amidst
vegetation not over 5 inches high.

Round-headed Thrift. *Armeria cephalotes.* Europe.
Evergreen perennial; 1 to 2 feet; pink; summer; seed
or division.—Banks, slopes and rocky places, associated
with Aquilegias and the finer and dwarfer perennials.

Lady Larpent's Plumbago. *Plumbago Larpentæ.*
China. Evergreen perennial; 1 foot; dark blue; summer; division or cuttings.—Banks, slopes or rocky
places, in any soil, amidst dwarf or prostrate plants.

THE VIRGINIAN POKE FAMILY.

Virginian Poke. *Phytolacca decandra.* Virginia.
Herbaceous perennial; 4 to 6 feet; purple; summer; division or seed.—Isolated tufts near wood walks, associated
with the largest and most vigorous herbaceous plants;
in copses, margins of plantations, &c. Worth planting
for the sake of its berries.

Kernel-like Phytolacca. *Phytolacca acinosa.* Northern India. Herbaceous perennial; 4 to 6 feet;
summer; division.—Similar positions to preceding.

THE RHUBARB FAMILY.

Alpine Persicaria. *Polygonum alpinum.* Switzerland. Herbaceous perennial; 2 to 3 feet; white;
summer; seed or division.—Woody places or copses.

Brown's Persicaria. *Polygonum Brunonis.* Northern India.   Evergreen perennial ; 6 to 9 inches ; pink ; summer ; division or seed.—Slopes, banks, or rocky places in any soil.

New Zealand Persicaria. *Muhlenbeckia complexa.* New Zealand.   Evergreen climbing shrub ; 2 to 4 feet ; yellowish ; summer ; cuttings.—Banks, slopes, or rocky places, among twining or trailing plants.

Siebold's Persicaria. *Polygonum Sieboldii.* Japan. Herbaceous perennial ; 4 to 6 feet ; yellowish green ; summer ; division or cuttings.—Isolated specimens near wood or pleasure-ground walks, but most striking in the former position ; or associated with the most vigorous herbaceous plants cultivated for the effect of their leaves or habit.

Whortleberry Persicaria. *Polygonum vaccinifolium.* Northern India.   Perennial ; 6½ to 12 inches ; pink ; summer ; division or cuttings.—Banks, slopes, rocky places, or margins of low shrubberies.

Emod's Rhubarb. *Rheum Emodi.* Nepal. Herbaceous perennial ; 4 to 6 feet ; purple ; summer ; division or seed.—Isolated tufts near wood-walks, or associated with other noble-leaved plants.   The tufts to be planted in the grass a few feet or yards from the margin of a plantation, in very deep, rich soil.

Palmate Rhubarb. *Rheum palmatum.* China. Herbaceous perennial ; 4 to 6 feet ; whitish ; summer ; division or seed.—Similar positions to the preceding, and associated with the Acanthuses, etc.

Common Rhubarb. *Rheum rhaponticum.* Asia.

Herbaceous perennial; 4 to 6 feet; white and green; summer; division or seed.—Similar positions to preceding.

Wave-leaved Rhubarb. *Rheum undulatum.* China. Herbaceous perennial; 4 to 6 feet; white and green; summer; division or seed.—Ditto.

THE BIRTHWORT FAMILY.

Great Birthwort. *Aristolochia Sipho.* North America. Deciduous climber; 10 to 30 feet; summer; division and cuttings.—A noble plant for covering arbours, banks, stumps of old trees, &c., also wigwam-like bowers, formed with branches of trees.

THE SPURGE FAMILY.

Cypress Spurge. *Euphorbia Cyparissias.* Southern Europe. Herbaceous perennial; 1 to 2 feet; yellow; summer; division or seed.—Slopes, banks, margins of copses, or rocky places, in any soil.

Glaucous Spurge. *Euphorbia Myrsinites.* Southern Europe. Evergreen perennial; 6 to 9 inches; yellow; summer; seed or cuttings.—Bare banks and rocky places in warm soils.

THE NETTLE FAMILY.

Rough Gunnera. *Gunnera scabra.* South America. Herbaceous perennial; 2 to 4 feet; summer; flowers inconspicuous; seed or division. — In warm and sheltered spots, in very deep and moist soil. Fine foliage.

THE ARROWHEAD FAMILY.

Broad-leaved Arrowhead. *Sagittaria latifolia plena.* North America. Aquatic; 1 to 2 feet; white; summer;

division.—In margins of ponds, rivers, boggy ground, or any position where there is mud or water.

THE ORCHID FAMILY.

Showy Lady's Slipper. *Cypripedium spectabile.* North America. Herbaceous perennial; 1 foot; white and rose; summer; division.—In places where boggy or deep, rich, moist, peaty or vegetable soil occurs. As it is rare it is better planted in a somewhat sheltered position, where it may escape injury from wind.

THE INDIAN SHOT FAMILY.

Mealy Thalia. *Thalia dealbata.* Carolina. Aquatic; 4 feet; blue; summer; division or seed.—Planted in the water near the margins of streams or lakes in deep soil; usually grown as a stove plant, but will be found to succeed well in the southern parts of this country.

THE FLOWER-DE-LUCE FAMILY.

Crested Flower-de-Luce. *Iris cristata.* North America. Herbaceous perennial; 3 to 6 inches; pale blue; summer; seed or division.—Low banks and rocky places, amidst very dwarf vegetation, in light sandy soil.

De Berg's Flower-de-Luce. *Iris De Bergii.* Belgian Hybrid. Herbaceous perennial; 2 to 3 feet; yellow and black; early summer; division.—Banks, margins of shrubberies, copses, groups, or in open glades, and associated with large and handsome herbaceous plants; thrives best in rich soil.

Pale Flower-de-Luce. *Iris flavescens.* Southern Europe. Herbaceous perennial; 2 to 3 feet; yellow; early summer; seed or division.—Similar positions and soil to preceding; not so ornamental.

Florentine Flower-de-Luce. *Iris florentina.* Southern Europe. Herbaceous perennial; 2 feet; white; early summer; seed or division.—Same soil and positions; an excellent plant for naturalization.

Common Flower-de-Luce. *Iris germanica.* Germany. Herbaceous perennial; 2 feet; blue; early summer; division or seed.—Same soil and positions as for I. De Bergii. There are many fine varieties.

Grass-leaved Flower-de-Luce. *Iris graminea.* Austria. Herbaceous perennial; 1 foot; purplish blue; early summer; division or seed.—Fringes of shrubberies or in rocky places, in ordinary soil.

Monnier's Flower de Luce. *Iris Monnieri.* Levant. Herbaceous perennial; 6 to 9 inches; yellow; early summer; division or seed.—In moist soil in glades and open parts of copses, or on fringes of shrubberies.

Tall Flower-de-Luce. *Iris ochroleuca.* Levant. Herbaceous perennial; 3 feet; creamy yellow; summer; seed or division. — Fringes of woods and copses or shrubberies, somewhat within the margin.

Pale Blue Flower-de-Luce. *Iris pallida.* Turkey. Herbaceous perennial; 3 feet; pale blue; early summer; division or seed.—Similar positions to those for the common Iris germanica.

Dwarf Flower-de-Luce. *Iris pumila.* Austria. Herbaceous perennial; 3 to 6 inches; purple; spring;

division.—Bare, low, rocky places, or level banks, always amidst dwarf vegetation. Thrives in ordinary soil, but best in peat.

Netted Flower-de-Luce. *Iris reticulata.* Iberia. Bulb; 6 inches; blue; spring; division.—Sunny spots on low warm banks in well-drained, rich, light soil.

Elder-scented Flower-de-Luce. *Iris sambucina.* Southern Europe. Herbaceous perennial; 3 feet; light blue; summer; division or seed.—Similar positions to those for Iris De Bergii.

Violet Flower-de-Luce. *Iris subbiflora.* Portugal. Herbaceous perennial; 2 feet; violet; summer; division or seed.—Similar positions to those for Iris germanica, but amidst dwarfer vegetation.

Variegated Flower-de-Luce. *Iris variegata.* Hungary. Herbaceous perennial; 2 to 3 feet; pale yellow; early summer; division or seed.—Similar soil and positions to those for Iris De Bergii, which is probably a variety of it.

Naked-flowered Crocus. *Crocus nudiflorus.* Southern Europe. Bulb; 6 inches; purple; autumn; division. —In the grass in glades, by wood-walks, &c. This plant is naturalized in several parts of England.

Showy Crocus. *Crocus speciosus.* Hungary. Bulb; 3 inches; purple; autumn; division.—In similar positions to preceding : also on banks and bare spots near fringes of shrubberies, &c., in sunny spots.

Susian Crocus. *Crocus susianus.* Turkey. Bulb; 3 to 4 inches; yellow; early spring; division.—Sunny banks, in short grass, in any soil.

Imperati's Crocus. *Crocus Imperatonius.* Italy.

Hardy bulb; 4 to 6 inches; lilac purple; very early spring.—In short grass not mown till rather late in the season, in warm, sunny spots where its early-flowering habit may be encouraged.

The common yellow and blue Crocuses, C. aureus and C. vernus may be naturalized with facility in grassy spots, as may the Scotch Crocus, C. biflorus, and indeed most of the other species.

THE DAFFODIL FAMILY.

Yellow Sternbergia. *Sternbergia lutea.* Southern Europe. Bulb; 6 to 9 inches; yellow; late in summer; division.—Bare places or low banks, well exposed to the sun, in gravelly and dry soil.

Atamasco Lily. *Zephyranthes Atamasco.* North America. Bulb; 6 to 9 inches; white; early summer; division or seed.—Low banks and rocky places, or here and there amidst very dwarf shrubs.

Hoop-petticoat Daffodil. *Narcissus Bulbocodium.* Southern Europe. Bulb; 6 inches; yellow; spring; division.—Low sunny banks and slopes, near pleasure-ground walks.

Two-coloured Daffodil. *Narcissus bicolor.* Southern Europe. Bulb; 1 foot; white and yellow; early summer; division.—In grassy places near wood-walks.

Peerless Daffodil. *Narcissus incomparabilis.* Portugal. Bulb; 1 foot; yellow; early summer; division.— On banks or slopes, in glades, or almost any position in any soil.

Great Daffodil. *Narcissus major.* Spain. Bulb;

1 foot; yellow; spring; division.—In tufts by pleasure-ground or wood-walks, in any soil.

Small Daffodil. *Narcissus minor.* Spain. Bulb; 2 to 4 inches; yellow; spring; division.—Associated with the dwarfest bulbous plants, on warm banks, and amidst very dwarf vegetation, in light sandy soil.

Sweet-scented Daffodil. *Narcissus odorus.* Southern Europe. Bulb; 1 foot; yellow; early summer; division. —Fringes of shrubberies, low sunny banks, in tufts in glades, and by wood walks, in any soil.

Poet's Daffodil. *Narcissus poeticus.* Southern Europe. Bulb; 1 foot; white; early summer; division.— Banks, by pleasure-ground or wood walks, margins of shrubberies, or in glades; in fact, in almost any position, and in almost any soil.

Plaited Snowdrop. *Galanthus plicatus.* Crimea. Bulb; 6 inches; white; winter; division. — In any position where the common Snowdrop succeeds.

Late Snowflake. *Leucojum Hernandezii.* Europe. Bulb; 1 to 2 feet; white; early summer; division.—In tufts in grassy places near wood and pleasure-ground walks, associated with the larger Daffodils.

THE DAY LILY FAMILY.

Two-rowed Day Lily. *Hemerocallis disticha.* China. Perennial; 2 feet; deep orange; summer; division.— Banks, slopes, tufts in glades, or associated in almost any position, with the stronger herbaceous plants. Grows in almost any soil.

Yellow Day Lily. *Hemerocallis flava.* Siberia.

Perennial; 2 feet; yellow; summer; division.—Similar positions to the preceding.

Tawny Day Lily. *Hemerocallis fulva.* Levant 2 feet; tawny; summer; division.— Similar positions and soil to those for H. disticha.

Grass-leaved Day Lily. *Hemerocallis graminea.* Siberia. Herbaceous perennial; 1 foot; yellow; summer; division.—Banks, slopes, or fringes of shrubberies, amidst very dwarf vegetation.

White Funkia. *Funkia grandiflora.* Japan. Herbaceous perennial; 1 to 1½ feet; white; summer; division or seed.—Warm banks, or sunny nooks amidst dwarf shrubs in good sandy loam.

Siebold's Funkia. *Funkia Sieboldiana.* Japan. Herbaceous perennial; 1 foot; whitish; summer; seed or division.—Isolated tufts near pleasure-ground walks. Thrives best in deep sandy peat.

Showy Tritoma. *Tritoma Uvaria.* Cape of Good Hope. Perennial; 3 to 4 feet; scarlet and yellow; late in summer; division. — Isolated tufts in glades, or associated with groups of the nobler autumnal-flowering herbaceous plants by wood walks.

THE LILY OF THE VALLEY FAMILY.

French Solomon's Seal. *Polygonatum intermedium.* A fine plant resembling our own Solomon's Seal, and suitable for similar positions. I have only seen it with M. Boreau in the Botanical Garden at Angers.

Large-flowered Trillium. *Trillium grandiflorum.* North America. Tuber; 6 to 9 inches; white; early

I

summer; division.—In moist, depressed, perfectly sheltered and shady nooks, in rich, deep vegetable soil.

Rose-coloured Lapageria. *Lapageria rosea.* Chili.
Climber; 6 to 15 feet; red; summer; cuttings or seed.
—I have been informed that this plant has been successfully established in the south of England in peat beds,
and allowed to twine among shrubs.

THE ASPHODEL FAMILY.

Yellow Asphodel. *Asphodelus luteus.* Sicily. 2 feet;
yellow; summer; division or seed.—In copses, margins
of shrubberies, and associated with medium-sized herbaceous plants.

St. Bruno's Lily. *Czackia (Paradisia) Liliastrum.*
Europe. Herbaceous perennial; 1 to 2 feet; white;
early summer; division or seed.—In glades, near wood
walks, or in the rougher parts of the pleasure-grounds.
Till plentiful it would perhaps be best tried in favourable
spots on low unmown banks in sandy loam.

Narbonne Star of Bethlehem. *Ornithogalum narbonnense.* Southern Europe. Bulb; 1 to 2 feet; white;
summer; division.—On low banks, associated with the
finest daffodils.

Pyramidal Star of Bethlehem. *Ornithogalum pyra-
midale.* Spain. Bulb; 1 to 2 feet; white; summer;
division.—Fringes of shrubberies, here and there among
dwarf shrubs, and associated with the finer Irises and the
dwarfer Lilies.

Byzantine Squill. *Scilla amœna.* Levant. Bulb;
3 to 6 inches; blue; spring; division.—On low, warm

banks, or in a sunny aspect amidst the grass, near the fringe of a shrubbery, or on slopes.

Two-leaved Squill. *Scilla bifolia.* Europe. Bulb; 4 inches; blue; spring; division.—On low banks, in light soil; also associated with the crocuses, snowdrops, &c. here and there in the grass. It is very hardy and free, but it is better to encourage its early-flowering tendencies by placing it in a warm and sunny position.

Bell-flowered Squill. *Scilla campanulata.* Spain. Bulb; 1 foot; blue; early summer; division.—Fringes of shrubberies, or in almost any position.

Spreading Squill. *Scilla patula.* Southern Europe. Bulb; 1 foot; blue; early summer; division.—Will thrive in positions in which the bluebell is found, but it is best first established on a sheltered sunny bank.

Italian Squill. *Scilla italica.* Italy. Bulb; 6 to 12 inches; blue; early summer; division.—Should be first established on warm banks or slopes, amidst very dwarf vegetation in light soil.

Siberian Squill. *Scilla sibirica.* Siberia. Bulb; 3 to 4 inches; blue; spring; division.—Bare sunny banks, slopes, rocky places, &c. Thrives well in good sandy loam, amongst very dwarf vegetation.

Amethyst Hyacinth. *Hyacinthus amethystinus.* Southern Europe. Bulb; ¾ foot; blue; early summer; division.—Similar positions to preceding.

Grape Hyacinth. *Muscari botryoides.* Italy. Bulb; 6 inches; blue; early summer; division.—Fringes of low shrubberies, and also on banks, associated with the finer squills and the amethyst hyacinth.

Musk Hyacinth. *Muscari moschatum.* Levant Bulb; 6 to 12 inches; brown and yellow; early summer; division.—Bare sunny banks near walks; only desirable for its odour, as the flowers are almost inconspicuous.

Large Yellow Allium. *Allium Moly.* Southern Europe. Bulb; 1 foot; yellow; summer; seed or division.—Fringes of copses, banks, and woody places. Associated with Ramsons (A. Ursinum).

Neapolitan Allium. *Allium neapolitanum.* Italy. Bulb; 1 foot; white; summer; seed or division.—Associated with the finer daffodils on banks, fringes of shrubberies, and in rough rocky places.

Broussonet's Asparagus. *Asparagus Broussoneti.* Canaries. Perennial; 3 to 10 feet; white and green; early summer; seed or division.—In shrubberies and copses, or among the most rampant climbing plants. Runs speedily up the stem of a dead tree or any similar object.

Esculent Camassia. *Camassia esculenta.* Columbia. Bulb; 1 foot; light blue; summer; division or seed.—On low banks, margins of shrubberies, and associated with the taller bulbous plants in light, well-drained, and warm soil.

New Zealand Flax. *Phormium tenax.* New Zealand. Evergreen perennial; 5 to 6 feet; buff; summer; division or seed.—This plant is tender in many parts of the country, but thrives very well in mild districts in the south and west of England and Ireland, in deep rich soil, and in shady or half-shady spots, in woods and copses. Where it grows well, isolated tufts of it have a fine effect near wood walks.

THE LILY FAMILY.

Thready Adam's Needle. *Yucca filamentosa.*
North America. Evergreen herb; 2 to 3 feet; white
and green; summer or autumn; suckers. — Margins
of shrubberies, on banks or rough rockwork, near cascades, &c.; and also, as it is a free-flowering kind, for
association with the nobler herbaceous plants.

Flaccid Adam's Needle. *Yucca flaccida.* North
America. Evergreen herb; 3 to 5 feet; whitish; summer or autumn; suckers.—Similar uses and positions to
the preceding; and, as it flowers more freely and regu-
larly, it is very valuable as a flowering plant.

Adam's Needle. *Yucca gloriosa.* America. Evergreen shrub; 4 to 6 feet; white; summer; suckers.—
Isolated specimens by wood walks, on very rough, rocky
ground, near cascades, &c.; or grouped with tall herbaceous plants of striking foliage and habit.

Recurved Adam's Needle. *Yucca recurva.* North
America. Evergreen shrub; 3 to 4 feet; white;
summer; suckers.—Similar uses and positions to preceding.

Gesner's Tulip. *Tulipa Gesneriana.* Levant. Bulb;
1½ feet; striped; spring; division.—This, or any of its
varieties, might be naturalized with the finer daffodils, or
hardy bulbs in warm good soil.

Sweet-scented Tulip. *Tulipa suaveolens.* Southern
Europe. Bulb; 6 to 9 inches; red and yellow; spring;
division.—Warm and sunny banks where there is not a
rank vegetation.

Rough-stemmed Tulip. *Tulipa scabriscapa.* Italy. Bulb; 1 to 1½ feet; red and yellow; spring; division. —Associated with daffodils and finer hardy bulbs, on banks, fringes of shrubberies, &c. Many of our earlier, or bedding tulips, are nearly related to this plant.

Crown Imperial. *Fritillaria imperialis.* Persia. Bulb; 3 feet; yellow; spring; division.—Fringes of woods and shrubberies; sometimes in isolated tufts at some distance from the margin of wood walks.

White Lily. *Lilium candidum.* Levant. Bulb; 3 to 4 feet; white; summer; division.—Fringes of shrubberies, single plants dotted, here and there, in beds of Rhododendrons or low shrubs, where they make a fine show when in flower, and do not afterwards, as the leaves die down, interfere with the general effect of the other plants.

Trumpet Lily. *Lilium longiflorum.* China. Bulb; 1 to 2 feet; white; summer; division.—The outer fringes of plantations, and beds of shrubs, or low banks and slopes, in light and sandy but deep soil.

Orange Lily. *Lilium croceum.* Italy. Bulb; 3 to 4 feet; yellow; summer; division.—Isolated tufts in glades, and similar positions to those for the White Lily. Like all Lilies, best in deep soil.

Golden-rayed Lily. *Lilium auratum.* Japan. Bulb; 3 to 4 feet; white, with golden stripes and dark-brown spots; summer; division.—Single plants of this fine Lily placed here and there among choice shrubs in good soil will produce a very fine effect.

American Turban Lily. *Lilium superbum.* North

America. Bulb; 4 feet; light orange; summer; division.—In shrubberies or the shade of trees planted in peat soil.

THE COLCHICUM FAMILY.

Spring Bulbocodium. *Bulbocodium vernum.* Spain. Bulb; 3 to 4 inches; purple; spring; division.—With the dwarfest and earliest Crocuses and other spring bulbs on warm banks and slopes.

Checkered-flowered Meadow Saffron. *Colchicum variegatum.* Greece. Bulb; 3 to 6 inches; lilac; autumn; division.—With the Crocuses and dwarfer bulbs, or in any position where our native Colchicum will grow.

Blistered Helonias. *Helonias bullata.* North America. Tuberous perennial; 1 foot; purple; early summer; division.—In boggy or moist sandy soil, associated with Cypripedium spectabile, Rhexia virginica, and other first-rate bog-plants.

White Veratrum. *Veratrum album.* Europe. Perennial; 4 to 5 feet; white; summer; division.—Associated with fine-foliaged plants, or by itself on fringes of shrubberies, or in rocky places.

THE PONTEDERIA FAMILY.

Heart-leaved Pontederia. *Pontederia cordata.* North America. Aquatic; 2 feet; blue; summer; division.— In water, within a few feet of the bank of lakes, ponds, &c., associated with such plants as the Sagittarias and the flowering rush.

THE SPIDERWORT FAMILY.

Virginian Spiderwort. *Tradescantia virginica.* North America. Herbaceous perennial; 1 to 2 feet; blue; summer; seed or division.—Naturalization on the low fringes of shrubberies, on low banks, rocky places, or slopes where there is a heavy soil or otherwise, or on level, wet, cold ground, or in bogs. There are several varieties, all equally free and hardy in the worst and coldest ground.

THE AROID FAMILY.

Aquatic Orontium. *Orontium aquaticum.* North America. Aquatic; 10 to 18 inches; striped; early summer; seed or division.—Margins of ponds and fountain basins, in rich soil, and kept apart from coarse aquatic plants.

Ethiopian Lily. *Calla æthiopica.* South Africa. Aquatic; 2 to 3 feet; white; early summer; seed or division.—Ponds and fountain basins in warm districts.

Marsh Calla. *Calla palustris.* North America. Aquatic; 6 to 9 inches; white; summer; division.—Wet muddy soil on margins of ponds, in shallow water, or in bogs and marshy places.

Common Dragon Arum. *Arum Dracunculus.* Southern Europe. Tuber; 1 to 1½ feet; brown; summer; division.—Among shrubs on the sunny side of walls or banks, preferring a warm sandy soil.

Hairy-sheathed Arum. *Arum crinitum.* Minorca. 1 to 2 feet; brown; spring; division.—Warm spots on sunny banks, where it may not be overrun by coarse plants. This remarkable kind is at present rare in cultivation, and deserves a favourable position.

THE GRASS FAMILY.

Bulbous Panicum. *Panicum bulbosum.* South America. Perennial; 3 to 4½ feet; seed or division.— Low banks and unmown spots in wild parts of the pleasure-ground, or associated with the finer perennials or with plants grown for the beauty of their leaves or habit.

Elegant Panicum. *Panicum capillare.* America. Annual; 1½ to 2 feet; summer; seed.—Banks, slopes, fringes of shrubberies, or with strong annuals in almost any position. If the Pampas, Arundo Donax, A. conspicua, and other ornamental grasses were grouped in a glade, this and the following might be associated with them, though less vigorous, in deep, rich, and well-drained soil.

Twiggy Panic Grass. *Panicum virgatum.* North America. Perennial; 3 to 6 feet; summer; seed or division.—Similar uses to preceding. This is an elegant plant, intermediate in stature between the small and very large ornamental grasses.

Feather Grass. *Stipa pennata.* Southern Europe. Perennial; 1 to 2 feet; summer; seed or division.— Banks, rocky places, or unmown spots near wood walks.

Great Reed Grass. *Arundo Donax.* Southern Europe. Perennial; 8 to 10 feet; summer; division.— In isolated tufts in glades open and sunny, but sheltered; or grouped with other noble grasses like the Pampas

Grass or Arundo conspicua, in a deep sandy soil, on a free dry bottom.

Pampas Grass. *Gynerium argenteum.* S. America. 6 to 9 feet; autumn; seed or division.—It is needless to recommend this noble plant, which is suited for and will adorn almost any position, but attains greatest perfection in warm and sunny, but sheltered spots, where its long leaves may not suffer from winds. It loves a rich, deep, sandy loam. Where wild groups of the nobler herbaceous plants are formed, this should always be a conspicuous object.

Graceful Bamboo. *Arundinaria falcata.* China, Japan. Evergreen shrub; 6 to 20 feet; summer; division only. —Positions similar to those for Arundo Donax. In southern and western districts, particularly near the sea, this plant thrives vigorously, attaining great size and beauty of habit.

Greyish Bamboo. *Bambusa viridi-glaucescens.* China. 7 to 12 feet; division.—A fine hardy bamboo, which will thrive in similar positions and districts to those for A. falcata, growing even more rapidly than that plant.

Briza-like Brome Grass. *Bromus brizæformis.* Sicily. Annual; 1 to 1½ feet; summer; seed.—A beautiful and graceful grass, which it would be well worth while to try and establish on rather bare, warm banks, or slopes in good warm soil.

Quaking Grass. *Briza maxima.* Southern Europe. Annual; 1 to 2 feet; summer; seed.—Might be naturalized with Bromus brizæformis.

PART III.

——

SELECTIONS

OF

HARDY EXOTIC PLANTS

FOR

Naturalization in Various Positions.

SELECTIONS

FOR NATURALIZATION.

———

As it is desirable to know how to procure as well as
how to select the best kinds, a few words on the first
subject may not be amiss here.

A very important point is the getting of a stock of
plants to begin with. In country or other places where
many good old border flowers remain in the cottage
garden, many species may be collected there. A
series of nursery beds should be formed in some by-
place in which such subjects could be increased to any
desired degree. Free-growing spring-flowers like Aubrietia,
Alyssum, and Iberis may be multiplied to any extent by
division or cuttings. Numbers of kinds may be raised
from seed sown rather thinly in drills, in nursery beds in
the open air. The catalogues should be searched every
Spring for suitable and novel subjects. The best time
for sowing is the Spring, but any time during the Summer
will do. Many perennials and bulbs must be bought in
nurseries and increased as well as may be in nursery beds.
As to soil, &c., the best way is to avoid the trouble of
preparing it except for specially interesting plants. The
great point is to adapt the plant to the soil—in peaty
places to place plants that thrive in peat, in clay soils

those that thrive in clays, and so on. My list of selections will to some extent help the reader in this way, while the soil suited for every plant has been, so far as I could advise, indicated under each in Part II.

A Selection of Plants for Naturalization in Places devoid of any but dwarf vegetation, on bare banks, etc.

Helleborus niger	Tunica Saxifraga
„ olympicus	Saponaria ocymoides
„ atrorubens	Silene alpestris
Aquilegia, in var.	„ Schafta
Pæonia tenuifolia	Cerastium Biebersteinii
Epimedium pinnatum	„ grandiflorum
„ alpinum	„ tomentosum
Dielytra eximia	Linum alpinum
Cheiranthus alpinus	„ arboreum
Arabis albida	„ flavum
Aubrietia, in var.	Geranium Wallichianum
Alyssum saxatile	„ striatum
Odontarrhena Carsinum	„ cinereum
Iberis corifolia	Oxalis floribunda
„ sempervirens	Genista sagittalis
„ correæfolia	Anthyllis montana
Thlaspi latifolium	Astragalus monspessulanus
Æthionema coridifolium	Coronilla varia
Helianthemum, in var.	Hedysarum obscurum
Viola cornuta	Vicia argentea
„ cucullata	Orobus vernus
Gypsophila repens	„ lathyroides
Dianthus neglectus	Waldsteinia trifolia

Potentilla calabra
Œnothera speciosa
„ missouriensis
„ taraxacifolia
Sedum dentatum
„ kamtschaticum
„ Sieboldii
„ spectabile
„ spurium
Sempervivum calcareum
„ hirtum
„ montanum
„ soboliferum
„ sedoides
Saxifraga Aizoon
„ cordifolia
„ crassifolia
„ crustata
„ longifolia
„ Cotyledon
„ rosularis
Astrantia major
Dondia Epipactis
Athamanta Matthioli
Cornus canadensis
Scabiosa caucasica
Hieracium aurantiacum
Doronicum caucasicum
Aster alpinus
Tussilago fragrans
Achillea aurea

Symphyandra pendula
Campanula carpatica
„ fragilis
„ garganica
„ cæspitosa
Erica carnea
Menziesia empetriformis
Gaultheria procumbens
Vinca herbacea
Gentiana acaulis
Phlox stolonifera
„ subulata
Lithospermum prostratum
Pulmonaria grandiflora
„ mollis
Myosotis dissitiflora
Physalis Alkekengi
Pentstemon procerus
Veronica austriaca
„ candida
„ taurica
Teucrium Chamædrys
Ajuga genevensis
Dracocephalum argunense
„ austriacum
Scutellaria alpina
Prunella grandiflora
Stachys lanata
Zietenia lavandulæfolia
Zapania nodiflora
Dodecatheon Meadia

Primula, in var.	Euphorbia Cyparissias
Acantholimon glumaceum	Iris cristata
Armeria cephalotes	„ graminea
Plumbago Larpentæ	„ pumila
Polygonum Brunonis	„ reticulata
„ vaccinifolium	„ nudicaulis

Plants of vigorous Habit for Naturalization.

Trollius altaicus	Thermopsis barbata
„ napellifolius	Spiræa, in var.
Thalictrum aquilegifolium	Astilbe rivularis
Delphinium, in var.	„ rubra
Aconitum, in var.	Molopospermum cicutarium
Pæonia, in var.	Ferula communis
Papaver orientale	„ glauca
Macleya cordata	„ tingitana
Datisca cannabina	„ sulcata
Crambe cordifolia	Peucedanum involucratum
Althæa ficifolia	„ longifolium
„ nudiflora	Heracleum eminens
„ taurinensis	„ flavescens
Lavatera Olbia	„ giganteum
Hibiscus militaris	Morina longifolia
„ Moscheutos	Dipsacus laciniatus
„ roseus	Mulgedium Plumieri
„ palustris	Alfredia cernua
Melianthus major	Onopordon tauricum
Galega officinalis	Cynara Scolymus
Lathyrus latifolius	Centaurea babylonica
Lupinus polyphyllus	Echinops bannaticus

Echinops exaltatus	Acanthus spinosissimus
„ ruthenicus	Phytolacca acinosa
„ purpureus	„ decandra
Aster elegans	Polygonum Sieboldii
„ Novi Belgii	Rheum Emodi
„ Novæ Angliæ	„ palmatum
„ pyrenæus	Gunnera scabra
„ ericoides	Thalia dealbata
Eupatorium purpureum	Tritoma Uvaria
Telekia cordifolia	Achillea Eupatorium
Helianthus angustifolius	Arundo Donax
„ multiflorus	„ conspicua
„ orgyalis	Gynerium argenteum
Harpalium rigidum	Bambusa falcata
Silphium perfoliatum	Elymus arenarius
Campanula pyramidalis	Veratrum album
Asclepias Cornuti	Yucca filamentosa
„ Douglasii	„ flaccida
Phlox, tall herbaceous vars.	„ recurva
Verbascum Chaixii	„ gloriosa
Physostegia imbricata	„ Treculeana
„ speciosa	„ aloifolia
Acanthus latifolius	Peucedanum ruthenicum
„ spinosus	Carlina acanthifolia

A Selection of Plants with large or graceful foliage suitable for naturalization.

Acanthus, several species	Morina longifolia
Asclepias syriaca	Polygonum cuspidatum
Statice latifolia	Rheum Emodi, & other spec.

K

Euphorbia Cyparissias

Datisca cannabina

Veratrum album

Tritomas, in var.

Thalictrum fœtidum

Crambe cordifolia

Althæa taurinensis

Geranium anemonæfolium

Melianthus major   [cus

Dimorphanthus mandchuri-

Elymus arenarius

Bambusa, several species

Arundinaria falcata

Yucca, several species

Verbascum Chaixii

Aralia spinosa

Spiræa Aruncus

„   venusta

Astilbe rivularis

„   rubra

Eryngium, several species

Ferula, several species

Seseli,   „

Chamærops excelsa

Hibiscus roseus

Rhus glabra laciniata

Artemisia annua

Phytolacca decandra

Centaurea babylonica

Lobelia Tupa

Peucedanum ruthenicum

Heracleum, several species

Aralia japonica

„   edulis

Macleya cordata

Panicum bulbosum

„   virgatum

Kochia scoparia

Dipsacus laciniatus

Alfredia cernua

Cynara horrida

„   Scolymus

Carlina acanthifolia

Telekia cordifolia

Echinops exaltatus

„   ruthenicus

Helianthus argyrophyllus

„   orgyalis

„   multiflorus

Gunnera scabra

Salvia argentea

Arundo Donax

„   conspicua

Gynerium argenteum

Silybum eburneum

„   marianum

Onopordon Acanthium

„   arabicum

A Selection of Hardy Plants of fine habit, that may be raised from Seed.

Among suitable hardy plants that may be raised from seed, the following are offered in recent seed catalogues :—

Acanthus latifolius

   „   mollis

   „   spinosus

Artemisia annua

Astilbe rivularis

Campanula pyramidalis

Cannabis gigantea

Carlina acanthifolia

Datura ceratocaula

Echinops, several species

Eryngium bromeliæfolium

   „   campestre

   „   cœlestinum

   „   giganteum

Ferula communis

   „   tingitana.

Geranium anemonæfolium

Gunnera scabra

Gynerium argenteum

Helianthus argyrophyllus

   „   orgyalis

Heracleum eminens

   „   giganteum

   „   platytænium

Kochia scoparia

Lobelia Tupa

Morina longifolia

Onopordon arabicum

   „   tauricum

Centaurea babylonica

Panicum, several species

Phytolacca decandra

Salvia argentea

Silybum marianum

   „   eburneum

Statice latifolia

Tritomas, in var.

Yucca, several species

Plants for Hedgebanks and Bushy Places.

Clematis in great var.

Thalictrum aquilegifolium

Anemone japonica & vars.

Delphinium, in var.

Aconitum, in var.

Macleya cordata

Hibiscus militaris	Campanula, in var.
Kitaibelia vitifolia	Calystegia dahurica
Tropæolum speciosum	„ pubescens
Baptisia australis	Verbascum Chaixii
Coronilla varia	Pentstemon barbatus
Galega officinalis	Veronica, tall kinds in var.
Astragalus ponticus	Phlomis Russelliana
Lathyrus grandiflorus	„ herba-venti
„ rotundifolius	Physostegia speciosa
„ latifolius	„ virginica
„ „ albus	Dracocephalum, in var.
Lupinus polyphyllus	Acanthus spinosus
Rubus biflorus	Statice latifolia
Œnothera macrocarpa	Phytolacca decandra
„ Lamarckiana	Boussingaultia baselloides
Astilbe rivularis	Aristolochia Sipho
Eryngium amethystinum	Asparagus Broussoneti
Molopospermum cicutarium	Vitis, in var.
Ferula, in var.	Honeysuckles, in var.
Morina longifolia	Ivies, in var.

Trailers, Climbers, etc.

The selection of plants to cover bowers, trellises, railings, old trees, stumps, rootwork, &c. suitably is an important matter, particularly as the plants fitted for these purposes are equally useful for rough rockwork, precipitous banks, flanks of rustic bridges, river-banks, ruins natural or artificial, covering cottages or outhouses, and many other uses in garden, pleasure-ground, or wilderness.

Vitis æstivalis
„ amooriensis
„ cordifolia
„ heterophylla variegata
„ Isabella
„ Labrusca
„ laciniosa
„; riparia
„ Sieboldii
„ vinifera apiifolia
„ vulpina
Hedera (the Ivy; all the named varieties, both green and variegated)
Aristolochia Sipho
„ tomentosa
Clematis azurea-grandiflora
„ campaniflora
„ elliptica
„ Flammula
„ florida
„ „ plena
„ „ Standishi
„ Fortunei
„ Francofurtensis
„ Hendersoni
„ insulensis
„ Jackmani
„ lanuginosa
„ montana

Clematis nivea
„ patens Amelia
„ „ Helena
„ „ insignis
„ „ Louisa
„ „ monstrosa
„ „ Sophia
„ „ violacea
„ pubescens
„ rubro-violacea
„ Shillingii
„ Sieboldii
„ tubulosa
„ Viticella
„ „ alba
„ „ venosa
Calystegia dahurica
„ pubescens plena
Wistaria sinensis
Asparagus Broussoneti
Periploca græca
Hablitzia tamnoides
Boussingaultia baselloides
Menispermum canadense
„ virginicum
Cissus orientalis
„ pubescens
Ampelopsis bipinnata
„ cordata
„ hederacea

Ampelopsis tricuspidata	Lonicera Caprifolium
Jasminum nudiflorum	,, confusa
,, officinale	,, flava
,, revolutum	,, japonica
Passiflora cærulea	,, Periclymenum

Spring and Early Summer Flowers for naturalization.

Anemone alpina	Epimedium alpinum
,, ,, sulphurea	Papaver croceum
,, apennina	,, bracteatum
,, blanda	,, orientale
,, Coronaria	Dielytra eximia
,, fulgens	,, spectabilis
,, Hepatica	Corydalis capnoides
,, ranunculoides	,, lutea
,, trifolia	Cheiranthus alpinus
Ranunculus aconitifolius	,, Cheiri
,, amplexicaulis	Arabis albida
,, montanus	Aubrietia deltoidea
Helleborus niger	Alyssum saxatile
,, olympicus	Draba aizoides
Eranthis hyemalis	Iberis corifolia
Aquilegia alpina	,, sempervirens
,, canadensis	,, correæfolia
,, cærulea	Viola biflora
Pæonia albiflora	,, cornuta
,, officinalis	Dianthus neglectus
,, tenuifolia	Saponaria ocymoides
Epimedium pinnatum	Silene alpestris

Arenaria balearica

„ montana

Ononis fruticosa

Vicia argentea

Orobus flaccidus

„ cyaneus

„ lathyroides

„ variegatus

„ vernus

Centranthus ruber

Centaurea montana

Doronicum caucasicum

Thlaspi latifolium

Hesperis matronalis

Erica carnea

Vinca major

Gentiana acaulis

Phlox reptans

„ subulata

Lithospermum prostratum

Pulmonaria grandiflora

„ mollis

Symphytum bohemicum

„ caucasicum

Myosotis dissitiflora

Omphalodes verna

Verbascum Chaixii

Dodecatheon Jeffreyi

„ Meadia

Cyclamen europæum

Cyclamen hederæfolium

Soldanella alpina

Primula Auricula

„ ciliata

„ nivalis

„ Palinuri

„ sikkimensis

Iris amœna

„ cristata

„ De Bergii

„ flavescens

„ florentina

„ germanica

„ graminea

„ ochroleuca

„ pallida

„ sambucina

„ sub-biflora

„ variegata

Crocus aureus

„ speciosus

„ versicolor

„ susianus

Narcissus angustifolius

„ Bulbocodium

„ bicolor

„ incomparabilis

„ major

„ montanus

„ odorus

Narcissus poeticus	Hyacinthus amethystinus
Galanthus plicatus	Muscari botryoides
Leucojum pulchellum	„ moschatum
Paradisia Liliastrum	Allium neapolitanum
Ornithogalum umbellatum	„ ciliatum
Scilla amœna	Tulipa Gesneriana
„ bifolia	„ suaveolens
„ campanulata	„ scabriscapa
„ patula	Fritillaria imperialis
„ italica	Bulbocodium vernum
„ sibirica	Helonias bullata

Plants for Naturalization beneath Specimen Trees on Lawns, etc.

Where, as is frequently the case, the branches of trees, both evergreen and deciduous, sweep the turf—and this, as a rule, they should be allowed to do in nearly all cases where they are planted in purely ornamental grounds— a great number of pretty Spring flowers may be naturalized beneath the branches, where they thrive without attention. It is chiefly in the case of deciduous trees that this could be done ; but even in the case of conifers and evergreens some graceful objects might be dotted beneath the outermost points of their lower branches. However, it is the specimen deciduous tree that offers us the best opportunities in this way. We know that a great number of our Spring flowers and hardy bulbs mature their foliage and go to rest early in the year. They require light and sun in Spring, which they obtain abundantly under the deciduous tree ; they have time to flower and develop

their leaves under it before the foliage of the tree appears;
then, as the Summer heats approach, they are gradually
overshadowed by a cool canopy, and go to rest undisturbed ; but the leaves of the trees once fallen, they soon
begin to appear again and cover the ground with beauty.

An example or two will perhaps explain the matter
more fully. Take the case of, say, a spreading old specimen of the handsome Weeping Mountain Elm. Scatter
a few tufts of the Winter Aconite beneath it, and leave
them alone. In a very few years they will have covered
the ground ; every year afterwards they will spread a
golden carpet beneath the tree ; and when it fades there
will be no eyesore from decaying leaves as there would
be on a border—no necessity for replacing the plants
with others ; the tree puts forth its leaves, covering the
ground till Autumn, and in early Spring we again see our
little friend in all the vigour of his glossy leaves. In this
way this pretty Spring flower may be seen to much greater
advantage, in a much more pleasing position than in the
ordinary way of putting it in patches and rings in beds or
borders, and with a tithe of the trouble. There are many
other subjects of which the same is true. We have only
to imagine this done in a variety of cases to see to what
a beautiful and novel result it would lead. Given the
bright blue Apennine Anemone under one tree, the
Snowflake under another, the delicately toned Triteleia
under another, and so on, we should have a Spring
garden of the most beautiful kind. Of course the same
thing could be carried out under the branches of a grove
as well as of specimen trees. Very attractive mixed

plantations might be made by dotting tall subjects like the large Jonquil (Narcissus odorus) among dwarf spreading subjects like the Anemone, and also by mixing dwarf subjects of various colours : diversely coloured varieties of the same Anemone, for example, would look very attractive.

Omitting the various pretty British plants that would thrive in the positions indicated—these are not likely to be unknown to the reader interested in such matters—and confining myself to dwarf, hardy exotic flowers alone, the following are selected as among the most suitable for such arrangements as that just described, with some little attention as to the season of flowering and the kind of soil required by some rather uncommon species. A late-flowering kind, for example, should be planted under late-leafing trees, or towards the points of their branches, so that they might not be obscured by the leaves of the tree before perfecting their flowers.

Anemone angulosa
　　„　　apennina
　　„　　blanda
　　„　　Coronaria
　　„　　fulgens
　　„　　Hepatica
　　„　　stellata
　　„　　sylvestris
　　„　　trifolia
Arum italicum
Bulbocodium vernum

Corydalis solida
　　„　　tuberosa
Crocus Imperati
　　„　　biflorus
　　„　　reticulatus
　　„　　versicolor
Cyclamen hederæfolium
Eranthis hyemalis
Erythronium Dens-canis
Ficaria grandiflora
Galanthus plicatus

Iris reticulata

Muscari botryoides, and vars.

  „   moschatum

Narcissus, in var.

Puschkinia scilloides

Sanguinaria canadensis

Scilla bifolia

  „ sibirica

  „ campanulata

Sisyrinchium grandiflorum

Trillium grandiflorum

Tulipa, in var.

Plants for very moist rich Soils.

Althæa, in var.

Aphyllanthes monspeliensis

Astilbe rivularis

Aralia edulis

  „ nudicaulis

Artemisia, in var.

Asclepias Cornuti

Asphodelus ramosus

Aster, in var.

Baptisia exaltata

Butomus umbellatus

Calla palustris

Caltha palustris fl. pl.

Chrysobactron Hookeri

Campanula glomerata

Convallaria multiflora

Colchicum, in var.

Crinum capense

Cypripedium spectabile

Datisca cannabina

Echinops, in var.

Elymus, in var.

Epilobium, in var.

Eupatorium, in var.

Ficaria grandiflora

Galax aphylla

Galega officinalis

Gentiana asclepiadea

Gunnera scabra

Gynerium argenteum

Helianthus multiflorus pl.

    „   orgyalis

    „   rigidus

Helonias bullata

Hemerocallis, in var.

Heracleum, in var.

Iris, the beardless kinds in variety.

Juncus effusus spiralis

    „   „ variegatus

Liatris, in var.

Lythrum (roseum super-bum)
Mimulus, in var.
Molopospermum cicuta-rium
Mulgedium Plumieri
Myosotis dissitiflora
Narcissus, stronger kinds.
Nierembergia rivularis
Œnothera, in var.
Omphalodes verna
Onopordon, in var.
Pancratium illyricum
Parnassia caroliniana
Phlomis herba-venti
„   Russelliana

Physostegia speciosa
Phytolacca decandra
Rudbeckia hirta
Ranunculus amplexicaulis
„   parnassifolius
Sanguinaria canadensis
Sparaxis pulcherrima
Solidago, in var.
Statice latifolia
Swertia perennis
Telekia speciosa
Thalictrum, in var.
Trollius, do.
Vaccinium, do.
Veratrum, do.

Plants suited for Peat Soil.

Alstrœmeria, in var.
Andromeda, in var.
Azalea amœna
Bryanthus erectus
Calandrinia umbellata
Calluna, in var.
Cassiope, in var.
Chimaphila maculata
Chrysobactron Hookeri
Coptis trifoliata
Cornus canadensis
Cypripedium spectabile

Dentaria laciniata
Daphne Cneorum
Dryas octopetala
Epigœa repens
Epimedium, in var.
Erica, in var.
Funkia Sieboldii
„   grandiflora
Galax aphylla
Gaultheria procumbens
Gentians, in var.
Helonias bullata

Iris nudicaulis, pumila, and vars.

Jeffersonia diphylla

Lycopodium dendroideum

Leiophyllum buxifolium

Linnæa borealis

Menziesia, in var.

Parnassia caroliniana

Podophyllum peltatum

     ,,   Emodi

Polygala Chamæbuxus

Pyrola, in var.

Ramondia pyrenaica

Rhododendron, small kinds

Schizostylis coccinea

Sarracenia purpurea

Sisyrinchium grandiflorum

Spigelia marilandica

Trientalis europæa

Trillium grandiflorum

Vaccinium, in var.

Zephyranthes Atamasco

     ,,   candida

Lilium superbum

Plants suited for Calcareous or Chalky Soil.

Adenophora, in var.

Æthionema cordifolium

Anemone, in var.

Alyssum saxatile

Anthyllis montana

Antirrhinum rupestre

Cistus, in var.

Cheiranthus, in var.

Campanula, in var.

Carduus eriophorus

Cerastium, in var.

Coronilla, in var.

Dorycnium sericeum

Dianthus, in var.

Echium, in var.

Erodium, in var.

Genista, in var.

Geum, in var.

Geranium, in var.

Glaucium Fischeri

Gypsophila, in var.

Hedysarum, in var.

Helianthemum, in var.

Hemerocallis, in var.

Lunaria biennis

Lupinus polyphyllus

Onobrychis, in var.

Ononis, in var.

Ophrys, in var.

Othonna cheirifolia

Phlomis, in var.

Prunella grandiflora

Santolina, in var.
Saponaria ocymoides
Saxifraga (the encrusted and the large-leaved kinds.)
Scabiosa, in var.
Sempervivum, in var.
Sedum, in var.
Stokesia cyanea
Symphytum, in var.
Thermopsis fabacea
Thymus, in var.

Trachelium cæruleum
Trifolium alpinum
Triteleia uniflora
Tunica Saxifraga
Vesicaria utriculata
Vicia, in var.
Vittadenia triloba
Waldsteinia trifoliata
„ geoides
Zietenia lavandulæfolia
Pyrethrum Tchihatchewi
„ roseum

Plants suited for Dry and Gravelly Soil.

Achillæa, in var.
Æthionema cordifolium
Agrostemma coronaria
Alyssum saxatile
Antennaria dioica
Anthyllis montana
Antirrhinum rupestre
Arabis albida
Aubrietia, in var.
Armeria cephalotes
Artemisia, in var.
Cerastium, in var.
Carlina acanthifolia
Cheiranthus, in var.
Chrysopsis mariana
Cistus, in var.

Corydalis, in var.
Dianthus, in var.
Dracocephalum, in var.
Dielytra eximia
Dorycnium sericeum
Echium, in var.
Erinus alpinus
Erodium, in var.
Eryngium, in var.
Euphorbia Myrsinites
Fumaria, in var.
Geranium, in var.
Gypsophila, in var.
Helianthemum, in var.
Helichrysum arenarium
Hypericum, in var.

Iberis, in var.

Jasione perennis

Lavandula Spica

Linaria purpurea

Linum, in var.

Lithospermum prostratum

Lupinus polyphyllus

Modiola geranioides

Narcissus, in var.

Nepeta Mussinii

Onobrychis, in var.

Ononis, in var.

Opuntia Rafinesquiana

Ornithogalum, in var.

Paronychia serpyllifolia

Plumbago Larpentæ

Polygonum vaccinifolium

Pyrethrum Tchihatchewi

Bahia lanata

Reseda odorata

Santolina, in var.

Scabiosa, in var.

Sedum, in great var.

Sempervivum, in great var.

Saponaria ocymoides

Stachys lanata

Sternbergia lutea

Teucrium Chamædrys

Thlaspi latifolium

Thymus, in var.

Trachelium, in var.

Tussilago fragrans

   ,,   Farfara variegata

Umbilicus chrysanthus

Verbascum, in var.

Vesicaria utriculata

Plants for Fringes of Cascades, etc.

Iberis, in var.

Helianthemum, in var.

Genista prostrata

   ,,   sagittalis

Daphne Cneorum

Polygonum Brunonis

   ,,   vaccinifolium

Santolina Chamæcyparissus

   ,,   incana

Ivies, in var.

Cistus, in var.

Rhododendron hirsutum

   ,,   ferrugineum

Arabis albida

Alyssum saxatile

Lithospermum prostratum

Saponaria ocymoides

Phlox subulata

Saxifraga, in var. particularly the large-leaved ones

Pentstemon, several vars.

Vinca major

   „   minor

Vinca herbacea

Œnothera macrocarpa

   „   taraxacifolia

Selection of Alpine and Rock Plants for Growing on Old Walls, Ruins, very Stony Banks, etc.

Corydalis lutea

Cheiranthus Cheiri

   „   „ pl. in var.

Arabis albida

   „ arenosa

   „ lucida variegata

   „ petræa

   „ blepharophylla

Aubrietia, all the vars.

Hutchinsia petræa

Vesicaria utriculata

Schivereckia podolica

Alyssum montanum

   „ saxatile

   „ spinosum

Koniga maritima

Petrocallis pyrenaica

Draba aizoides

   „ bœotica

Ionopsidion acaule (north side of old walls)

Thlaspi alpestre

Iberis, in var.

Reseda odorata

Helianthemum, in var.

Gypsophila muralis

   „ prostrata

Tunica Saxifraga

Dianthus cæsius

   „ deltoides

   „ monspessulanus

   „ petræus

Saponaria ocymoides

Silene acaulis (moist walls, to be first carefully planted in a chink)

   „ alpestris

   „ rupestris

Silene Schafta

Lychnis alpina

   „ lapponica

Sagina procumbens pleno

Arenaria balearica

   „ cæspitosa

   „ ciliata

   „ graminifolia

   „ montana

   „ verna

Linum alpinum

Malva campanulata (ruins)

Erodium romanum (old walls)

„ Reichardii

Ononis alba

Astragalus monspessulanus

Coronilla minima

„ varia

Acæna Novæ Zealandiæ (moist mossy walls)

Cotyledon Umbilicus

Umbilicus chrysanthus

Sedum acre

„ „ variegatum

„ Aizoon

„ album

„ anglicum

„ brevifolium

„ cæruleum

„ dasyphyllum

„ elegans

„ Ewersii

„ farinosum

„ hispanicum

„ kamtschaticum

„ multiceps

„ pulchrum

„ sempervivoides

„ sexangulare

Sedum sexfidum

„ spurium

Sempervivum arachnoideum

„ arenarium

„ calcareum

„ globiferum

„ Heuffelli

„ hirtum

„ montanum

„ piliferum

„ tectorum

Saxifraga bryoides

„ cæsia

„ crustata

„ cuscutæformis

„ diapensioides

„ Hostii

„ intacta

„ lingulata

„ longifolia

„ pectinata

„ pulchella

„ retusa

„ rosularis

„ Rocheliana

„ Rhei

„ sarmentosa

Asperula cynanchica

Centranthus ruber

L

Centranthus albus	Veronica saxatilis
,, coccineus	Iris germanica and vars.
Santolina incana	,, pumila
Achillea tomentosa	Polypodium vulgare
Symphyandra pendula	Adiantum Capillus-Veneris
Campanula Barrelieri	(on moist warm walls)
,, fragilis	Asplenium Adiantum - ni -
,, garganica	grum
,, cæspitosa	,, fontanum
,, ,, alba	,, septentrionale
,, rotundifolia	,, Ruta-muraria
Antirrhinum rupestre	,, germanicum
,, majus	,, lanceolatum
,, Orontium	,, Trichomanes,
Linaria Cymbalaria	and vars.
,, ,, alba	,, viride
,, vulgaris	Ceterach officinarum
Erinus alpinus	Matthiola tristis
Veronica fruticulosa	

A Selection of Alpine and Rock Plants for Naturalization.

Anemone, in var.	Alyssum saxatile
Helleborus niger	Draba aizoides
Cheiranthus alpinus	,, bœotica
Arabis albida	Iberis corifolia
Aubrietia deltoidea	,, sempervirens
., purpurea	,, correæfolia
Alyssum montanum	Æthionema coridifolium

Æthionema saxatile

Helianthemum, in var.

Cistus, in var.

Polygala Chamæbuxus

Gypsophila repens

     „   prostrata

Dianthus, in var.

Tunica Saxifraga

Saponaria ocymoides

Silene alpestris

    „   Pumilio

    „   Schafta

    „   Elizabethæ

Spergula pilifera

Cerastium, in var.

Erodium Manescavi

Oxalis floribunda

Astragalus monspessulanus

Fragaria indica

Potentilla calabra

Œnothera taraxacifolia

    „   marginata

Calandrinia umbellata

Sedum, in var.

Sempervivum, in var.

Saxifraga, in var.

Hydrocotyle bonariensis

Dondia Epipactis

Linnæa borealis

Hieracium aurantiacum

Helichrysum arenarium

Othonna cheirifolia

Santolina, in var.

Achillea tomentosa

Erica carnea

Menziesia empetriformis

Gaultheria procumbens

Gentiana acaulis

Phlox stolonifera

    „   subulata

Convolvulus lineatus

Lithospermum prostratum

Myosotis azorica

    „   dissitiflora

Omphalodes verna

Linaria alpina

Antirrhinum rupestre

Pentstemon procerus

Erinus alpinus

Wulfenia carinthiaca

Veronica candida

Thymus corsicus

Zietenia lavandulæfolia

Zapania nodiflora

Soldanella alpina

Primula, in var.

Androsace Chamæjasme

    „   lanuginosa

Aretia Vitaliana

Acantholimon glumaceum

Armeria cephalotes	Polygonum vaccinifolium
Plumbago Larpentæ	Euphorbia Myrsinites
Polygonum Brunonis	

Selections of Alpine and Rock Plants with Prostrate or Drooping Habit, suited for placing so that they may Droop over the Brows of Rocks, and like Positions.

Arabis albida	Ononis arvensis albus
,, procurrens	Lotus corniculatus
Aubrietia, in var.	,, ,, fl. pl.
Alyssum saxatile	Astragalus monspessulanus
Iberis corifolia	Coronilla varia
,, Tenoreana	Vicia argentea
Helianthemum, many	Orobus roseus
kinds	Fragaria indica
Gypsophilas, several	Potentilla alpestris
Dianthus deltoides, and	,, calabra
others	,, Hopwoodiana
Tunica Saxifraga	,, M'Nabiana
Saponaria ocymoides	,, verna, and nu-
Cerastium Biebersteinii	merous vars and hybrids
,, grandiflorum	Œnothera acaulis
,, tomentosum	,, macrocarpa
Callirhoe involucrata	,, taraxacifolia
,, pedata	Sedum spurium
Tropæolum speciosum	,, kamtschaticum
,, polyphyllum	,, reflexum
Genista prostrata	,, Sieboldii
,, tinctoria	,, sempervivoides

Saxifraga, hypnoides and vars.

„ oppositifolia, and vars.

Linnæa borealis

Galium verum

Scabiosa graminifolia

Diotis maritima

Artemisia argentea

Campanula Barrelieri

„ carpatica

„ „ alba

„ fragilis

„ „ hirsuta

„ garganica

„ muralis

Erica carnea

Epigæa repens

Phlox subulata

„ reptans

Convolvulus mauritanicus

Lithospermum prostratum

Antirrhinum rupestre

Linaria Cymbalaria

Veronica taurica

Thymus lanuginosus

Thymus Serpyllum, white var.

Zietenia lavandulæfolia

Dracocephalum argunense

Zapania nodiflora

Plumbago Larpentæ

Lysimachia Nummularia

Polygonum vaccinifolium

Euphorbia Myrsinites

Salix lanata

„ reticulata

Empetrum nigrum

Polygonum complexum

Boussingaultia baselloides

Medicago falcata

Lathyrus grandiflorus

„ latifolius

„ „ albus

Vicia Cracca

Calystegia dahurica

„ pubescens

Vinca major

„ minor

„ herbacea

Clematises, the new varieties of the lanuginosa section

*List of Alpine and Rock Shrubs, etc., suitable for
Naturalization in Bare, Rocky, and Peaty Places,
associated with the finer Herbaceous Plants.*

Iberis, in var.
Helianthemum, in var.
Cistus, in var.
Polygala Chamæbuxus
Hypericum
Genista tinctoria
    ,,   sagittalis
    ,,   prostrata
Hedera, variegated and
  other curious vars.
Othonna cheirifolia
Erica carnea, and all hardy
  species and vars.
Arbutus Uva-ursi
Pernettya mucronata
Gaultheria procumbens
Andromeda hypnoides
    ,,   fastigiata
    ,,   tetragona
Menziesia cærulea
    ,,   empetriformis
    ,,   polifolia, and
vars.
Daphne Cneorum
Lithospermum prostratum
Polygonum vaccinifolium

Veronica saxatilis
    ,,   taurica
Euphorbia Myrsinites
Salix lanata
    ,,   reticulata
    ,,   serpyllifolia
Empetrum nigrum
Santolina Chamæcyparis-
  sias
    ,,   incana
Euonymus radicans varie-
  gata
Rhododendron hirsutum
    ,,   ferrugineum
    ,,   Chamæcistus,
  and others
Bryanthus erectus
Azalea amœna
Epigæa repens
Skimmia, in var.
Vaccinium Myrtillus
    ,,   macrocarpum
    ,,   Oxycoccos
    ,,   Vitis-idæa
    ,,   uliginosum
Juniperus squamata

A Selection of Annual Plants for Naturalization.

Papaver somniferum
Eschscholtzia californica
Platystemon californicum
Matthiola annua
    „   bicornis
Arabis arenosa
Alyssum maritimum
Ionopsidium acaule
Iberis coronaria
    „   umbellata
Malcolmia maritima
Erysimum Peroffskianum
Heliophila araboides
Gypsophila elegans
Saponaria calabrica
Silene Armeria
Viscaria oculata
Malope trifida
Tropæolum majus
Limnanthes Douglasii
Ononis viscosa
Œnothera odorata
Godetia Lindleyana
    „   rubicunda
    „   tenella

Clarkia elegans
    „   pulchella
Eucharidium   concinnum
    grandiflorum
Amberboa moschata
    „   odorata
Helianthus annuus
Dimorphotheca pluvialis
Gilia capitata
    „   tricolor
Collomia coccinea
Leptosiphon androsaceus
    „   densiflorus
Nicandra physaloides
Collinsia bicolor
    „   verna
Dracocephalum nutans
    „   moldavicum
Blitum capitatum
Polygonum orientale
Panicum capillare
Bromus brizæformis
Briza maxima
    „   gracilis
Agrostis nebulosa

A Selection of Biennial Plants for Naturalization.

Matthiola, in var.

Lunaria biennis

Hesperis matronalis

Erysimum asperum

Silene pendula

Hedysarum coronarium

Œnothera Jamesi

Œnothera Lamarckiana

Dipsacus laciniatus

Silybum eburneum

Onopordum illyricum

Campanula Medium

    ,,   ,,   rosea

Verbascum phlomoides

Ornamental Grasses for Naturalization.

Agrostis nebulosa

Briza maxima

Brizopyrum siculum

Bromus brizæformis

Hordeum jubatum

Panicum virgatum

    ,,   bulbosum

    ,,   capillare

Arundo conspicua

Arundo Donax

    ,,   ,,   variegata

Erianthus Ravennæ

Gynerium argenteum, and vars.

Polypogon monspeliensis

Stipa gigantea

    ,,   pennata

Milium multiflorum

Aquatic Plants for Naturalization.

Nuphar advena

Nymphæa Kalmiana

    ,,   odorata

Calla palustris

Pontederia cordata

Aponogeton distachyon

Orontium aquaticum

Trapa natans

Hardy Bulbs for Naturalization.

Allium Mcly	Hyacinthus amethystinus
„ fragrans	Iris, in great var.
„ neapolitanum	Leucojum vernum
„ paradoxum	Lilium, in var.
„ roseum	Merendera Bulbocodium
Brodiæa congesta	Muscari, in var.
Bulbocodium vernum	Narcissus, in great var.
Camassia esculenta	Ornithogalum, in var.
Crinum capense	Scilla, in var.
Crocus, in great var.	Sparaxis pulcherrima
Colchicum, in var.	Sternbergia lutea
Cyclamen, in var.	Trichonema ramiflorum
Erythronium Dens-canis	Triteleia uniflora
Fritillaria, in var.	Tulipa, in var.
Galanthus plicatus	Zephyranthes Atamasco
Gladiolus communis	„ candida

List of Plants for Naturalization in Lawns and other Grassy Places that are frequently mown.

This must of necessity be a limited list—being confined to subjects that will grow and flower very early in the season, and not form tufts or foliage large enough to injure the turf. Even with these it will be desirable to refrain from rolling or cutting the Grass as early as usual. For this and like reasons this is by no means so desirable as other ways which I recommend, and which can be carried out without check of any kind, and without interfering with anything except indeed the monotonous and uninteresting surfaces now seen in every pleasure ground.

Galanthus plicatus	Anemone blanda
„ nivalis	Narcissus minor
Leucojum vernum	Erodium Reichardi
Scilla bifolia, in var.	Sagina glabra
„ sibirica	Crocus, in var.

Plants for Dotting over Grass in Spots seldom Mown, or Mown very late in the Season.

Bulbocodium vernum	Dianthus deltoides
Colchicum, in var.	Erodium romanum
Cyclamen hederæfolium	Fumaria bulbosa
Galanthus plicatus	Helichrysum arenarium
„ nivalis	Iris reticulata
Leucojum vernum	Linum alpinum
Scilla bifolia	Narcissus minor
„ alba	„ bicolor
„ sibirica	„ Bulbocodium
„ italica	„ juncifolius
„ amœna	Sternbergia lutea
Anemone apennina	Zephyranthes candida
„ ranunculoides	Hyacinthus amethystinus
„ blanda	Merendera Bulbocodium
„ trifolia	Muscari, in var.
Antennaria dioica rosea	Trichonema ramiflorum
Anthyllis montana	Triteleia uniflora

PART IV.

———

THE GARDEN

OF

BRITISH WILD FLOWERS.

THE

GARDEN OF BRITISH WILD FLOWERS.

———

HOWEVER well people may be acquainted with the floral beauties of our fields and woods in Spring or Summer, few have any conception of the great number of really pretty flowers that may be selected from wild places in various parts of the British isles, and cultivated with success in a garden. Few of us, except working botanists, and they are sparsely scattered beings, have much notion of the great variety of beauty that may be culled from British flowers alone ; and as botanists very rarely *cultivate* wild flowers, they can quite as rarely select the kinds best suited for our gardens. Most of us have full opportunity of seeing the beauties of the fields and hedges ; not so many the mountain plants, and few such rare gems as Gentiana verna, which grows wild in Teesdale, and here and there on the western shores of Ireland, or the mountain Forget-me-not, a precious little dwarf alpine that is

found but rarely in Yorkshire and Scotland. It is
only by a careful selection from all classes of the
plants of the British isles that we can hope to
arrive at anything satisfactory in the way of a
" garden of British plants." I do not by this mean
a " scientific " or botanical arrangement of English
flowers, but a charming little hardy garden, or series
of beds filled exclusively with the better kinds of
our native plants, dotted here and there with our
native shrubs, and surrounded, if the situation
required it or admitted of it, with English
trees and shrubs, from the sweet gale to the
fragrant "May," or scarlet-berried Mountain Ash.
There is nothing difficult in the making of such
a garden, and I think its charms, to lovers of the
garden generally, would be very great. In it might
be exhibited the beauties of some of our prettiest
spring flowers, of not a few really showy plants and
neat dwarf shrubs, and of most of the charming
meadow flowers worth cultivating : while the
Orchids, which we generally have to seek with
some little patience, even in good plant districts,
might also be seen thriving in it. However, the
best plan of all is to scatter about our own wild
flowers in the wild and semi-wild places so often
before alluded to.

It is not only the curious and rare that may afford us interest among the vegetable natives of Britain ; among them are included things of a high order of beauty, that will flourish and keep their own ground without any watching or special preparation of the soil ; and even for the sake of selecting plants wherewith to embellish the margins of lakes, rivers, ponds, or beds of fountains in our parks, pleasure grounds, or gardens, the subject is worthy attention. For the rockwork, too, many of our wild flowers are well suited : and, if in making a special little arrangement for English plants, a bit of rockwork could be introduced, and near it, in the shade, a position for ferns, they would prove a useful addition. As regards the best way of growing them, or utilizing them in gardens generally, all will depend upon the size or nature of the place. Many of the plants may be grown with advantage in the small villa (or even the suburban) garden, and in a large one with plenty of space, a very pretty distinct feature might be made of them. In any part of the country where the soil or surface of the ground suits the habits of a variety of native plants, it would prove a most interesting employment to collect kinds not found in the neighbourhood, and naturalize them

therein; and wherever the natural rock crops
up in a picturesque way, a great deal of beauty
may be added to the place by planting these
rocky spots with wild flowers of a suitable nature.
There are hundreds of parks and grounds all over
the country that would grow to perfection the finer
wild flowers, in which noticeable kinds are not to be
seen, and when once a collection is obtained there
can be little difficulty in making good use of it.

Need we grow weeds to have a fair representa-
tion of beautiful British wild flowers? No such
thing! It will be my pleasant task to look over
the whole British flora with the reader, to tell him
where to find and how to grow the rarer kinds, and
to enumerate all that are ornamental; and in
doing so I shall have to name a great variety of
plants, but not one weedy subject I hope. In
the season of wild flowers, when many of us stray
into the fields, or on to the hills, to find many a
gem which I advise should be grown in the garden
instead of being made a mummy of, the more
beautiful British flowers will prove much more
delightful in wild and half-wild places near our
gardens, and scrambling over slopes and through
hedgerows, alive and full of change, than ever they
have done in the best herbarium.

So far as I am acquainted with the labours of British botanists or horticulturists, none of them have ever attempted a selection from our wild flowers as adapted for garden use. The botanist, as a rule, deals with things in a wild state only, and therefore the subject has never been thought of by him ; the horticulturist generally deals only with the useful or the conspicuously attractive, and has never thought of culling the higher beauties of our flora. But why should this be so ? " Botany," says Emerson, " is all names, not powers ;" and assuredly, if it does not lead us to a real enjoyment of our wild flowers, it is barely worthy of a better character. To flatten and dry a number of wild plants and leave them in dust and darkness is necessary for botanists, but it is not likely to cause any wide-spread human interest in such things ; and therefore I propose that we look through the list of British wild flowers and endeavour to rescue the subject from its present dry-as-dust character.

First it will be necessary to have a complete list of British wild flowers, which would be found in the index to Syme's, Bentham's, Babington's, or any other good book on our flora ; but best of all is a special list called the " London Catalogue of British Plants," which used to be published by Pamplin,

M

and is now, I think, published by Dulau, of Soho. This is particularly useful, because it gives a full list of all the species, and by means of numbers indicates their comparative prevalence. The compilers adopted Mr. Hewett Watson's division of Britain into a number of botanical districts, and after the name of each species a number is placed, which tells the number of districts in which that particular plant is found. Thus on the first page, "18" is placed after the name of the Marsh-marigold, indicating that this strong and beautiful herb is found in the eighteen districts, or, in other words, that it is very common. The Seakale (Crambe maritima) is put down as an inhabitant of twelve districts; and that pretty alpine plant the Yellow Draba (D. aizoides), is marked "I.," because it is only found wild in one district in Britain.

I think it very desirable that those who wish to work at the collecting and culture of wild flowers should provide themselves with one or more of these lists, simply for convenience sake, as on them may be at once marked the kinds we have or want; and I myself have found them very useful to effect exchanges, marking the species I had and could spare, and sending the list to friends in distant

parts of the country, who could collect many kinds
not in my neighbourhood, and who in their turn
required many things that I could collect plentifully
enough : thus we exchanged the Orchids of the
Surrey hills for the Alpines of the higher Scotch
mountains, and so on throughout the country. It
need scarcely be said that every student, cultivator,
or admirer of British plants should possess himself
of a manual by which he can identify the species,
and which will also probably hint where the species
may be found, and some other useful particulars.
Another valuable aid to some would be a " local
flora," a list of the plants growing in any particular
neighbourhood, or county, such, for instance, as the
" Flora of Reigate," Baines's " Flora of Yorkshire,"
and Mackay's " Flora Hibernica," or the recently
issued " Cybele Hibernica." It might prove interesting to some to cultivate the best of the local
plants, even if those from distant parts could not
be conveniently obtained.

So much for books ; we will next turn to the
plants themselves, beginning with the natural
order of Crowfoots, or Ranunculaceæ. This
is the order which brightens the moist hollows
in the Spring with the glittering gold of the
lesser Celandine, the meadows in May with Butter-

cups, when "those long mosses in the stream" begin to assume a livelier green, "and the wild Marsh Marigold shines like a fire in swamps and hollows grey." "Those long mosses in the stream" of "The Miller's Daughter" are simply some of the Water Crowfoots that silver over the pools with their pretty white cup-like blossoms in early Summer; and it is precisely the same brotherhood which burnishes our meadows and "stamps the season of Buttercups" with a glistening glory of colour not equalled by any tropical flowers I have ever met with. Now in going completely through the known species of British plants I propose to enumerate only those that are really worthy of garden culture, and certain to reward our trouble in gathering and planting them, and I do not recommend them from published plates and descriptions, but from actual experience in their culture.

The first plant named in books of British Plants is the Traveller's Joy (Clematis Vitalba), the well-known common clematis that streams over the trees, and falls in graceful folds from many a low tree in many parts of the south of England, and which is generally conspicuous enough in autumn from the heads of large feathery awns that abound on it at that season. It is of course well known

and deservedly employed as a garden plant, and
from its rapidity of growth nothing is better adapted
for quickly covering objects such as rough
mounds, &c. However, it may be most tastefully
used in the shrubbery or wilderness, and particularly so on the margin of a river, or water, where
the long streamers of its wiry branchlets look
effective and distinct at all times. It is the only
indigenous plant that affords any idea of the all-
embracing and interminable twiners or "bush
ropes," that run about in wild profusion in tropical
woods; and in some places in England it grows so
freely as to become a nuisance. The most natural
looking and prettiest bower I have ever seen was
formed by this plant running up a low oak and
falling down in thick festoons to the ground; by
pushing the twiners a little aside in the summer,
a most agreeable bower was at once formed.
There is scarcely any end to what may be done
with it in this way. The plant is to be had for a
trifle in most nurseries; it is abundantly wild in
the southern counties, and to be had in numerous
gardens.

Next we have the elegant lesser Thalictrum (T.
minus)—elegant, I say, because I have grown it, in
the open bed, so like the Maidenhair fern that some

of our most experienced cultivators were surprised
at the resemblance, and declared it to be every
whit as pretty for the open air as the Maidenhairs
are for the greenhouse; therefore I have recommended it as the "Maidenhair Fern" for the open
garden.   It only requires to be planted in ordinary
soil and left alone till it gets established. Then,
when the elegant leaflets unfold, all the grace and
distinctiveness of the fern before named will appear
in the open air, able to withstand all the sun that
can assail it in our clime, and charming for close
association with flowering plants. It is wild in
many parts of Britain, particularly in Scotland
and North-western England, and rather abundant
on the island of Ireland's Eye, near Dublin, in
many parts of the limestone districts of Clare and
Galway, and rises to a considerable elevation on
the mountains. It produces very insignificant
flowers, which should be pinched off immediately
when they are noticed, or rather, the flower-stem
should be pinched off the moment it begins to rise,
as all the beauty lies in the foliage, and therefore
the flowers must not be suffered to weaken it in
any way. It grows about a foot high, or perhaps
more in rich soil and when well established. There
are several other species natives of Britain, but

none of them nearly so well worthy of culture as this.

Next come the windflowers, or Anemones, four kinds, all good; two of them — A. nemorosa, the wood anemone, and A. apennina, the blue anemone—indispensable. The wood anemone is a charming thing, either in its wild or cultivated state, and besides the normal white variety there are a red and a bluish one, also a double white variety, very desirable, though not common. They grow in the open border, on rockwork, &c., quite as well as in the shade. As for the blue anemone, it is simply one of the loveliest spring flowers of any clime and should be in every garden, both in the borders and scattered thinly here and there in woods and shrubberies, so that it may become "naturalized." The flowers are freely produced, and of the loveliest blue. It is scarcely a true British flower, so to speak, its home being the south of Europe; but, having strayed into our wilds and plantations occasionally, it is now included in books on British plants, and may be easily obtained in most nurseries that grow spring flowers or herbaceous plants.

The Pasque anemone, or Pasque-flower, is an important native, bearing large flowers of a dull violet purple, silky outside. It is fond

of limestone pastures, and occurs in several districts in England, though it is wanting in Scotland and Ireland.   It is, however, a rare plant in
England, but may be seen occasionally in a garden
or nursery. Another kind, A. ranunculoides (yellow),
is a doubtful native found in one or two spots.
It is worth growing as a border plant, and must be
had from a nursery or garden, as it is not to be
found wild except in one locality.

So much for the Anemones, of which the first
two are the gems.   Adonis autumnalis is the very
pretty and conspicuous-flowered "pheasant's-eye,"
an annual plant of easy culture found occasionally
in corn-fields, and of which the seed is included in most lists under the name of Flos
Adonis.   It is singular rather than beautiful, and
though the flowers are very bright, it is not very
attractive.

The Ranunculuses, or crowfoots, begin with R.
aquatilis and its several varieties, and several other
species of Water Ranunculi with divided leaves.
Few gardens offer any facilities for cultivating these.
The most we can do is to introduce them to a pond
or stream in which they are not already found, or
add one of the long-leaved or rarer kinds to the
common kind or kinds ; but their home is in the

fresh stream, "hither, thither, idly playing," or in the lake or ornamental water, and therefore they hardly come among garden plants. I have tried to grow all the kinds I could get in a small pond; but the Canadian weed, or the common R. aquatilis, soon exterminated my rarities, and I was compelled to give it up, and look for the varied beauty of the water crowfoots in any passing stream. R. Ficaria is the pretty little shining-leaved yellow-flowered species which abounds in moist and shady land in spring, one of the earliest spring flowers that appears, and very common throughout Britain and many parts of Europe; but it is none the less beautiful because common, and although not fit for the garden, is very pretty in a woody waste in early spring. The roots are, to a great extent, masses of little cylindrical tubers, by which it is easily known.

R. Flammula (the spearwort) is a native of wet marshes and river-sides in all parts of Britain, and is well suited for planting by the side of a pond, brook, or ornamental water, though not so fine as the greater spearwort, R. Lingua, which is a noble, strong-growing kind, often growing two or three feet high, and bearing large, showy, yellow flowers. It is very fine near the margins of water, and is

rather freely scattered over the British isles, but
not common. These plants are of course only to
be collected in a wild state, though they are grown
in some botanic gardens. The others are what we
would mostly call wild field kinds, and are too apt
to become dangerous weeds to be admitted to the
garden. R. acris pleno and R. repens pleno are
double forms of the wild kinds, and well worth
growing, from their exceedingly pretty " bachelor's-
button" flowers, bright yellow, neat, and very
double. From being double they last longer in
flower than the single kinds, are well suited for use
among cut flowers, and are, in fact, very desirable
border plants. They must be had from a nursery,
or from a place where herbaceous flowers are
grown, though possibly they may be found wild
occasionally, though very rarely.

Then we have the large marsh marigold (Caltha
palustris), which makes such a glorious show in
spring along moist bottoms, or by river banks in
rich soil—notably on the left bank of the Thames
as you go to Kew, where, when there has been a
very high tide during the flowering season, I have
seen the ground for many feet under the water look
as if strewn with gold, in consequence of the water
having overflowed the banks and covered numbers

of these showy flowers when fully expanded. This is well worth introducing to the margin of all garden waters, or even to moist ground, where it is not already established, because it makes a truly fine spring-flowering plant. There is a double variety sold rather plentifully in Covent Garden in early summer, which is very desirable, bearing perfectly double flowers of large size, which, like the double Crowfoots last longer than the single blooms. Trollius europæus is the pretty Globe-flower, well worthy of a place from its clear yellow colour, pleasing outline, and sweetness. Not a common plant in England, but rather frequent in the north and west, from Wales to the Grampians, and in Ireland. It grows nicely as a border plant. That pretty and dwarf little spring flower, the Winter Aconite (Eranthis hyemalis), also belongs to this order, and is most worthy of culture. It is naturalized here and there, and may be had abundantly from any bulb merchant or grower of spring flowers.

The English hellebores are barely worth growing except in a botanic garden. The common columbine (Aquilegia vulgaris) is desirable, and often very pretty. It is not very common in the wild state, but undoubtedly a real native in several

counties of England. In one gorge on Helvellyn
I have found it ascend almost to the top of that
mountain, flowering beautifully in most inaccessible
sible spots ; it is rather common in gardens, and to
be had from the seedsmen. Delphinium Consolida
is somewhat frequent in the eastern counties ; it is
an annual, interesting and desirable where a full
collection is sought, but has hardly quality enough
for the choice selection. The common poisonous
aconite (A. Napellus) is rather an ornamental
native plant, though only locally distributed ; it is,
however, very common in gardens, where it should
be carefully isolated from any roots likely to be
used as food, in consequence of the frightfully
poisonous character of its roots.

The common Berberis vulgaris, which is rather
widely distributed, must not be forgotten among
British plants, for I doubt if there be a more
beautiful sight afforded by any shrub than by this
when draped over with its brightly-coloured racemes
of fruit, which are also so useful in the edible department.
partment. I remember having been more struck
with the beauty of several fine old bushes of this
plant at Frogmore than with any other shrubs in
the wide gardens there. There is a vulgar but
quite unfounded prejudice which charges the plant

with blighting crops, but it should be in every garden, and in a large place might be planted in the shrubbery or adjacent to the British collection. The queenly white Water-lily, so common in our rivers, should be seen in all garden waters, not thickly planted, but a single specimen or group here and there. It is most effective when one or a few good plants are seen alone on the water; then the flowers and leaves have full room to develop and float right regally; but when a dense crowd of water lilies are seen together, they are usually poorly developed, and crowd each other out. The effect is never half so beautiful as when—

Some scatter'd water-lily sails
Down where the shallower wave still tells its bubbling tales.

See how the author of " Childe Harold" chances inadvertently to note the beauty of the Water-lily when isolated, compared to what it is when choked together in a river bed or garden water. With it should be associated the yellow Water-lily (Nuphar lutea), and if the small and rare Nuphar Pumila can be had, so much the better.

Among the poppies, the only one really worth growing as a garden plant is the Welsh Poppy (Meconopsis cambrica), which grows so abundantly along the road sides in the lake district. It is a pretty perennial of a clear yellow, and thrives well at the bottoms of walls and such positions. Some might care to grow the large Opium Poppy (P. somniferum); its finer double varieties are doubtless very good, but these can scarcely be called British. The Horned Poppy of our sea shores is distinct and curious, and on that account might be grown in a garden; but it must be treated as an annual or biennial. Corydalis solida is a pleasing and dwarf spring flower, scarcely a native, or very rare; and the seakale, really ornamental when in flower, is well worthy of a place on a wild bank.

In the natural order Cruciferæ, Thlaspi alpestre (a pretty Alpine), Iberis amara (a fine white annual), Draba aizoides (a rare and beautiful Alpine), Koniga maritima, the sweet Alyssum, and Dentaria bulbifera, rare, pretty, and curious; Cardamine pratensis, the ladies' smock, and its double variety; Arabis petræa, a sweet dwarf alpine; the common wallflower, and the single rocket (Hesperis matronalis) will be found the most ornamental, and all of them are worth growing. Any Flora of the United Kingdom will tell their habitats. None of the mignonettes found in Britain are worthy of cultivation.

All the British Helianthemums or rock roses are worthy of a position on the rockwork, and the annual kind H. guttatum, is a singularly pretty thing, with black spots at the base of its clear yellow petals. Of the Violets, in addition to the sweet violet, which should be grown on a north aspect, V. lutea and V. tricolor will be found the most distinct and worthy of culture. The Droseras, or sun-dews, are very pretty, but cannot be long preserved in a garden; nor have I ever seen the pretty Polygalas cultivated with success. The very dwarf trailing Frankenia lævis (Sea Heath) runs over stones, and looks neat and mossy on a rockwork.

In the Pink tribe, the scarce, single, wild Carnation (D. Caryophyllus), D. plumarius, by some supposed to be the parent of the garden pinks, and D. cæsius, the Cheddar Pink, which does so nicely on an old wall or on rockwork, D. deltoides, the maiden pink, the common soapwort (Saponaria officinalis), the sea bladder-campion (Silene maritima), Silene acaulis, the beautiful little alpine that clothes our higher mountains, the corn cockle (Lychnis Githago), the Ragged Robin, and the alpine lychnis; the vernal sandwort (Arenaria verna), Arenaria ciliata, found on Ben Bulben, in Ireland, and Cerastium alpinum are the best, and

these are all worthy of culture. The last is as
shaggy as a Skye terrier, and does not grow more
than an inch high. I have found it thrive out of
doors in a garden near London, though people
generally treat it as a delicate alpine plant, and
grow it in frames.

A really ornamental species of Flax is not by
any means a common inmate of British gardens,
but a pretty species occurs in some of our eastern
counties, and may be seen in most botanic gardens
and some nurseries. This is Linum perenne, a
pretty blue-flowering, medium-sized border plant.
There is a pure white variety, which is fully equal
to the blue, or even better, because pure white
border flowers are not so plentiful. Both are quite
hardy and perennial, well suited for rockwork or
the most select mixed border. There is also a
rose-coloured variety, but whether the "rose" be
worthy of that name or not I cannot say, as I
have not flowered the plant. The Perennial
Flax, or any of its varieties, will be found to
thrive in any place where the grass is not mown
as well as on borders. Seed is offered by
various seedsmen, so that there need be no
difficulty in raising plants of this most desirable
British species. None of the other British Flaxes

are worthy of cultivation. The common flax is sometimes found wild, but it is not a true, or at all events is a very doubtful, native.

In the natural order Malvaceæ, we have several showy plants, but none particularly worthy of garden cultivation, except it be Lavatera arborea (the Sea Lavatera), which is sparsely distributed along the south and west coasts, and on the Bass Rock in the Frith of Forth. It is a plant of vigorous habit and noble leaves, which might be used with advantage in what is nowadays called the subtropical garden, or, indeed, in almost any position, for it is a plant of very distinct habit. It grows five or six feet high when in a favourable situation. The best of the mallows is the Musk Mallow (M. moschata), which bears a profusion of rather showy flowers in summer. It is not an uncommon English plant, and would not discredit the mixed border. The Marsh Mallow (Althæa officinalis) will of course be cultivated for other reasons than its beauty, which is not very striking. The Marsh Mallow is found in the south of England, but does not go far north, nor is it very common, whereas the common mallow is to be seen everywhere, except perhaps in the extreme north.

N

Among the Hypericums there is something to admire ; indeed, .nearly all of them possess some beauty, and might find a place among low shrubs ; but by far the best is H. calycinum, or " St. John's wort," a kind which is not perhaps truly British naturally, but which is to be seen in many gardens, and is now naturalized in several parts of England and Ireland in bushy places.   The very large and showy flowers of this species, combined with its dwarf and neat habit, make it fit for a place in any garden, and it is particularly adapted for rough rockwork, or will crawl away freely under and near trees, &c. ; though of course, like most things, it will best show its beauty when fully exposed to the sun and air.   It is a plant that can be had everywhere.

In the Geranium order there are a few pretty things for the garden—notably, G. pratense, G. sylvaticum, and G. sanguineum, with its fine variety G. lancastriense.   This variety was originally found in the Isle of Walney, in Lancashire, and some writers have made it a species under the name of G. lancastriense, but most good botanists now consider it a variety of G. sanguineum.   Both plants are well worth growing in a garden.   The latter is widely distributed in Britain, and yet is not very general. It is what might be called a local plant, while G. lancastriense must now be sought for in nurseries or botanic gardens. G. sanguineum makes a very pretty border plant of dwarf and compact habit. G. phæum is a species with flowers of a peculiar blackish colour, and is more curious than beautiful. It is wild in some parts of Westmoreland and Yorkshire, and is worth a place in a full collection from its distinctness, if nothing else.

The common Oxalis (O. Acetosella) is the prettiest among its British allies ; and a chaste little plant it is, too, when seen luxuriating in shady, woody places, along hedge-banks, &c. It cannot be cultivated to perfection fully exposed, but in all gardens where there is a little diversity, or any half-wild, shady spot, it might be introduced with advantage. Some say it is the shamrock of the ancient Irish, but they are certainly wrong. Established custom among the Irish during the experience of the oldest inhabitants, and everything that can be observed or gleaned, tend to point to the common trifolium as the true shamrock.

In the Pea tribe there are a few plants of great merit, and the first we meet with is the very pretty dwarf shrub Genista tinctoria, or Dyer's genista.

This is an exceedingly neat little shrub, very low
and dwarf, but vigorous in the profusion of its
bright yellow flowers.   It ought to be in every
garden, and would be equally at home on the select
rockwork, the border, or among very dwarf shrubs.
It is rather frequent in England, but rare in Scotland and Ireland.   It can be had from most shrub
nurseries.   Its two allies, G. pilosa and G. anglica,
are also neat and interesting little shrubs, and
though not so decidedly ornamental as the Dyer's
genista, they are well worth a place in an interesting collection of dwarf British shrubs.

Most people who admire wild flowers must
have been struck with the beauty of the common
Restharrow, which spreads such a sheet of delicate
colour over many a chalk cliff and sandy pasture
or roadside.   It bears garden culture willingly, and
is prettier when in flower than numbers of New
Holland plants, which require greenhouse pro-
tection and ceaseless expense to keep them alive
at all.   There is a smoother, taller, and more bushy
form of this sometimes admitted as a species, G.
antiquorum, which is also a very ornamental plant,
and well suited for the mixed border.   These
plants grow very freely from seed, and are of the
easiest culture.

Among the Medicagos there is a good deal of coarse vigour ; but one of them, while not lacking vigour, I have found a very lovely plant for large rockwork or for the mixed border. M. falcata has decumbent stems, and forms a dense, wide-spreading mass upon the ground, the whole plant being covered with yellow flowers. Now, if M. falcata be planted on a rough rockwork, or any other position from which it can let fall its luxuriant, low-lying growth, it will prove a most ornamental object, and is of an almost perennial duration and great hardiness. Found only in southern and eastern England. The other Medicks and their allies possess some beauty, but scarcely sufficient to warrant their garden culture, and all of them are inferior to M. falcata.

None of the Clovers or Trifoliums can be recommended for garden culture, because the most showy kinds are common in our fields ; and therefore whatever garden space we can spare for wild flowers had better be devoted to things we are not likely to meet with every day. Here again it may be said that Trifolium repens is the true shamrock, and has been so since the days of St. Patrick. Some say that it is of comparatively recent introduction to Ireland, but without either proof or probability on their side, as it reaches nearly as far north as the Arctic circle; and why it should avoid such a genial spot as the green isle, we are not informed. Though comparatively common, the lotus or bird's-foot trefoil is so thoroughly distinct and beautiful, that it must not be omitted in "The Garden of British Wild Flowers," flowering as it does nearly the whole summer, and keeping so dwarf and neat in habit. There are several forms. I know of no better plant for the front edge of the mixed border. The lady's fingers, or Anthyllis vulneraria, is rather a pretty thing found in chalky pastures and dry stony places in England, and often grown as a farm plant.

The three British kinds of Astragalus are worthy of cultivation, and still more so is the allied genus, Oxytropis. Both O. campestris and O. uralensis are neat dwarf plants, the foliage of the last being quite silvery, and its habit one of the neatest. The first is only found in one spot among the Clova mountains in Scotland; the second is rather common on the Scotch hills. Hippocrepis comosa is rather like the bird's-foot trefoil, both in habit and flower, and is well worth a place among the choice dwarfs. Not found in Scotland or Ireland, but rather abundant in some parts of England.

Of the several kinds of Vicia, or vetch, two at least are eminently worthy of culture—V. Cracca and V. sylvatica. The first of these makes a charming border plant if slightly supported on stakes when young, so that it may have hidden its supports by the time the flowers appear. I have grown this a perfect wide-spreading mass of bluish purple, and it is one of the most conspicuous of herbaceous plants. The other kind is of a climbing habit, but most elegant when seen running up the stems of young trees or over bushes. This is found in most woody hilly districts of Britain and Scotland, and V. Cracca is common everywhere in this country. Among British peas decidedly the best is the Sea Pea, Lathyrus maritimus, which makes a remarkably handsome plant in rich deep ground; and, indeed, its large bluish purple flowers make it attractive on any soil. It occurs on the coast of southern and eastern England, of Shetland, and of Kerry, in Ireland. The seeds are edible, and have been used ere now by the country people as food.

In the Rose order both the Spiræas will repay attention; certainly S. filipendula, which, in addition to its pretty flowers, has leaves cut somewhat after the fashion of a fern, and may indeed be used in that capacity in the flower garden : it would furnish somewhat of the effect of a Blechnum. The double variety is very desirable. It is found freely enough in England, but does not go far into Scotland ; nor is it recorded from Ireland. Dryas octopetala, a plant found on the limestone mountains of North England and Ireland, and abundantly in Scotland, makes a neat border plant in light free soil and where the air is pure. About Edinburgh I have noticed pretty edgings made of it in some of the nurseries. Very near London it does not seem to do well ; but in all cases it is worthy of a trial, being an interesting and distinct wild flower.

As for the blackberry, raspberry, dewberry, and cloudberry, many may desire to cultivate them in the shrubbery, and very interesting it is to observe the differences between some of the sub-species and varieties of blackberries, and the beauty, both in fruit and flower, of the family. The cloudberry can only be grown in a cold, wet, boggy soil, and is almost impossible of culture as a garden plant, except in moist and elevated spots. The dewberry, distinguished principally by the glaucous bloom on the fruit when ripe, is of easy culture. Of the Potentillas, P. rupestris, white-flowered, found on the Breiddin Hills in Montgomeryshire, and the large golden-yellow-flowered P. alpestris, found on the higher limestone mountains, are the best. P. fruticosa, found in the north of England and in Clare and Galway, in Ireland, makes a neat, free flowering bush; and the marsh potentilla (P. Comarum) will do well in boggy ground, if you have such, though it is more distinct than pretty. As to the wild roses, it is difficult to make any selection, because of their great interest. All the species and varieties that could be collected would surely prove of great interest in the shrubbery, as would all the British trees and shrubs of the Rose family.

Everybody at all familiar with our native plants knows the common Willow-herb (Epilobium angustifolium), so showy, and so apt to become a disagreeable weed in some places. But if properly placed in some out-of-the-way spot, where it cannot overrun or interfere with rarer and less vigorous plants, it becomes a real ornament, even when contrasted with the most showy of exotic herbs. Even the botanist, in describing it, says, "a handsome plant"—an expression very rarely used by gentlemen who write on English botany. Though very widely distributed over Britain, it is not what

would be called a common plant ; but in no case
can there be any difficulty in obtaining it. Planted
and allowed to have its own way in a shrubbery, or
any other position you care little about, it will
furnish a rich display with its purplish red flowers
in summer. Behind the late Sir Joseph Paxton's
fine house at Chatsworth, there is a little private
garden, and the shrubbery that encloses this ex-
hibits an abundance of the Willow-herb, planted
there by Paxton, who, though he enjoyed the
noblest tropical plants near at hand in the great
conservatory and Victoria Regia house, yet was
alive to the charms of this fine native plant.
There are many other kinds, but none of them so
worthy of culture as this.

The Evening Primrose (Ænothera biennis), deserves a place from its fragrance ; and, as it is apt
to go wild, it is as well to place it in some out-of-
the way spot, where it may be found when desired,
and yet not have an opportunity to become a weed.
I observe it has quite covered waste building
ground near Westminster. As for the Marestail,
it is an aquatic plant, in general outline somewhat
resembling the Equisetums, and suited for the
curious and interesting collection rather than the
ornamental ; it flourishes healthfully in a ditch,

margin of pond, or fountain basin, placed in a pot, which will prevent its running about too much.

Next we have the distinct and showy purple Loosestrife (Lythrum Salicaria), a ditch and marsh plant, abundant in many parts of Britain. There is a variety of this plant known in nurseries and gardens by the name of L. roseum superbum, which should be in every garden, whether the owner takes an interest in English plants or not. This, planted by the side of ornamental water, makes a splendid object, and is also a first-rate border plant. The colour of its long spikes of flowers is of the most charming character. So, whatever you do in British flowers, do not forget Lythrum roseum superbum, or, in more correct language, the fine rose-coloured variety of the common Loosestrife. It may now be had in the nurseries, and is used as a flower-garden plant in some parts of the North. It may be easily raised from seed, which is offered in some catalogues.

The common Herniary (Herniaria glabra) and Scleranthus perennis are two very dwarf green spreading plants, found in some of the southern and central counties of England, and which would furnish a neat Lycopodium-like effect on rockwork or anything of the kind. Their flowers are almost inconspicuous, but the habit is neat, and the tone refreshing.

Then we come to the Roseroot (Sedum Rhodiola) and the tribe of neat, pretty, and interesting Sedums, every one of which is worthy of a place on the rockwork or rocky bed in the "garden of British wild flowers"—from the common stonecrop, which grows on the thatch of cottages and abundantly in many parts of Britain, on walls and rocky places, to that little gem for a wall or rockwork, the thick-leaved Sedum dasyphyllum of the south of England. This last is perhaps not truly native in Britain, but can be readily had wherever collections of these plants are grown. The Roseroot is so called from the drying root-stock smelling like roses. The Orpine or Livelong (Sedum Telephium) is also a fine old plant of this order.

Grow the British sedums on a little slightly rocky or elevated bed, but they will do quite well on the fully exposed level ground ; only keep them free from weeds, or, from their diminutive size, they may become exterminated by these, or even by the common stonecrop, which usually makes a vigorous attempt to grow through and choke up the smaller members of its family. If you have any old walls or buildings try and establish a few of the smaller kinds on these ; and while it is very interesting to have rare plants established in such places, you will find that the timid and tenderer kinds will always survive on them ; whereas they may get cut off by the winter when on level ground or in pots. This is particularly true of the charming little Sedum dasyphyllum, which everybody having an old wall, or mossy old building of any kind, would do well to endeavour to establish, by putting a young plant in a suitable chink with a little sandy soil around it. Once it has seeded, in all probability the plant will become firmly established ; the seedlings raised on the wall are sure to live long and perpetuate themselves.

Not a few small and delicate plants that can hardly be preserved long in a garden in any other way may be grown on a wall. If you are a fern-grower, you will know how difficult it is to establish the little Wall Rue (Asplenium Ruta-muraria) in pots, pans, or any way in the hardy fernery ; but by taking a few of the spore-bearing little fronds, and putting a little of the " fern-seed" into the chinks of an old wall, you will soon establish it ; and in like manner it is quite possible to cultivate the Ceterach and the graceful Spleenwort (often

erroneously called the Maiden-hair), only that the
wall must be somewhat older and richer, so to
speak, to accommodate these than the Wall-rue.
Indeed, this little fern will grow on a wall that is in
perfect condition, as may be seen by any person
driving past Lord Mansfield's place at Highgate,
where the high garden-wall that runs for some distance parallel with the road running from Hampstead to Highgate is covered in its upper part with
this plant, and would be so lower down, and more
abundantly, were it not for the depredations of
plant-collectors. In a moister district, or on an
older wall, it would, of course, be far more luxuriant ; but the fact that you may establish it on a
sound wall is worth relating. Nothing could be
more interesting than to see an old wall covered
with ferns, draped here and there with Linaria, and
studded in spots with the Sedum above recommended for this purpose, or with others.

The Sedums are succeeded in the natural classi-
fication of British plants by the Saxifrages—
beautiful, most interesting, and very neat in habit,
like the Sedums in size, but distinct and even more
indispensable for the garden. First, there is the
Irish group of Saxifrages, the London Pride and
its varieties ; and the Killarney saxifrage, S.

Geum and its interesting varieties, both species very pretty for rockwork and borders.

Next we have the mountain S. stellaris and S. nivalis, and the yellow marsh S. Hirculus, and the yellow S. aizoides, which fringes the rills and streams on the hills and mountains in Scotland, and the north of England and Ireland, all interesting and worthy of a trial — but far surpassed by the purple Saxifraga oppositifolia, which opens its vivid purplish-rose flowers soon after the snow melts on its native rocks in the Scotch Highlands, and as far north, among the higher mountains of Europe, Russian and Central Asia, as the Arctic circle. It bears garden culture well, either as a pot plant in the cold frame or pit, on the rockwork, or in patches in the front of a border. In planting this it would be well to excavate holes a couple of feet deep, filling them again with a mixture of broken stone and earth, so that when the roots descended among these evaporation-preventing stones, they might find a good substitute for that moisture and that nutriment which they enjoy among the *débris* and in the chinks of their native rocks. The purple saxifrage should be planted in the full sun.

This caution is the more necessary in conse-

quence of nearly every person who grows these interesting dwarf plants, keeping them in a shady
position, in which they soon perish, or never look
such far-glistening ornaments as when grown in the
full sun, and supplied with a sufficiency of water.

The meadow Saxifraga granulata differs in most
respects from most of the other members of the
family that are in cultivation, and is worth growing ; its double variety, which may be seen in many
cottage gardens, is much used in some places for
the spring garden, and is in all respects a most desirable garden plant.   It flowers so abundantly
that the very leaves are hidden by the profusion of
rather large double flowers.   I have noticed it frequently in small cottage gardens in Surrey, and it
may now be had easily from the nurserymen.   It
is a pleasing and much admired subject for the
spring garden.

The dense green moss-like Saxifrages are a
most important group for the garden, in consequence of the fresh and living green which they
assume in winter, when everything else begins to
look lamentable and ragged — when the fallen
leaves rush by, driven by the wet gusts of autumn
—and when geraniums and all the fleeting flower-
garden plants are cut off by the frost.   They grow

on almost any soil or situation, and may be grown
with ease even in large towns, provided always
that they are fully exposed to the sun, and get a
few thorough waterings during very dry summers.
They are dotted over with white flowers in early
Summer, the stems of which should be cut off as
soon as the flowers perish; but to me their great
beauty is in Autumn, when they glisten into various
tints of the most refreshing green, and all through
the winter, when they remain in the same condition, or emerge from the deepest snows verdant as
leaves in June. S. hypnoides, abundant in Scotland, Wales, and northern England, with its varieties,
is our most important plant in this way; and S.
cæspitosa, found on some of the higher Scotch mountains, is nearly allied to it, and of nearly equal merit.
There is no necessity for going to the Scotch or
any other mountains for these mossy Saxifrages,
as they are grown a great deal here and there—may
be had from many nurseries—and seed is offered
in some catalogues.

Green is attractive to many people, especially
in winter, and to those whose eyes require refresh-
ment after severe mental exertion or sedentary
work—a very large class indeed, nowadays. In
towns it is difficult to get shrubs to retain

their verdure, in consequence of smut and other adverse influences ; in all places these mossy Saxifrages will afford it most attractively if planted on some borders near the window, or better still, on a rather flat-lying fringe of rockwork opposite them.   I have seen a gentle bank, facing the drawing-room window of a house, covered most effectively in this way, having it studded with a few "rocks," and then planting it with a variety of these mossy Saxifrages and a few other perpetually green, hardy dwarf plants. In winter it was most refreshing to look upon—more attractive than the evergreen shrubs beyond it.

Next we have the beautiful Grass of Parnassus (Parnassia palustris), a distinct and charming native plant, rather frequent in Britain in bogs and moist heaths. I have grown it very successfully in a small artificial bog, and still better in six-inch pots in peat soil, the pots being placed in a saucer of water during summer, and preserved in a cold frame in winter. It is, however, much better to "naturalize" it in moist grassy places than to grow it in this way.

The Spignel or Baldmoney (Meum athamanti-cum), which is found in the Scotch highlands, in Wales, and the north of England, and has most elegantly divided leaves, being very dwarf and neat in habit, is a most desirable border or rock plant. The flowering stems should be pinched out, as it is for the much-dissected leaves only that the plant is worthy of cultivation. In the whole of the umbelliferous order there is hardly another plant worthy of cultivation, if we except the Sea-holly (Eryngium maritimum), a striking subject, and the Sweet Cicely (Myrrhis odorata) an interesting old plant, often cultivated in old times and gardens for various uses; not a rare plant, but most plentiful in the hilly parts of the north of England. The rest of the order are best admired in their wild haunts, like a great many other British plants.

The Linnæa borealis is one of the prettiest and most distinct things among our native plants, and, moreover, highly attractive to all who know anything about botany, in consequence of its being named after the great master of natural science, Linnæus himself, who was very fond of this plant, which trails about so prettily in fir woods in the North. It is found, though rarely, in Scotland ; but being a favourite plant with many, may be purchased in many nurseries. The only question is, how to keep a plant so interesting and pretty ? To place it in the ordinary earth of our dry southern gardens would be a ready way of extinguishing it ; but by a little management it may be grown quite readily by anybody. I have grown it in three different ways. First in the open garden, planted in deep silvery peat, and covered with a hand-glass, rubbed over with a half-dry paint brush, to furnish the necessary amount of shade. In that way it did very well,—luxuriantly, in fact. The glass, nearly quite close at all times, preserved the desired moisture around the plant, and it never required any attention, except to remove weeds now and then.

Of course anybody can follow the same practice. As a painted handlight is not a very ornamental object, it would of course be better to place it in some shady or out-of-the-way spot. Such will also accord better with its character. Another equally successful way is to plant it in a moist, cool, shady, cold frame, such as you would use for bringing on a batch of young hardy ferns—the frame to face the north instead of the south, as is usually the case. By putting some peat and leaf-mould in the back of such a frame, and planting a nice little specimen or two of the Linnæa, I have had it nearly fill the frame. In a like kind of frame it may be grown to perfection in pots of peat, the peat to be kept very moist. In such, when it becomes well established, the graceful shoots hang in a mass over the pot, and then it may be removed for some time to the outside of a window on the shady side of a house, the pot being placed in a saucer constantly filled with water. Thus you may enjoy, even without leaving the house, a plant that any botanist would be grateful to you for growing, so much do botanists admire it, while it is at the same time pretty enough to ensure admiration from those unlearned in plants.

That the Heath family is likely to afford much interest I need hardly remind any person who has seen the wide spread of beauty on our heaths and mountains in summer or autumn. But of the variety of loveliness which exists among our native heaths few people have any idea : not even the sportsman or botanist, who continually wanders over their native wilds, or the plant collector, with a quick eye for everything attractive or noble in the way of a plant. The species themselves are of course very beautiful ; but from time to time sports have appeared amongst them which nurserymen have preserved ; and thus, where you see a good collection of these, the variety of gay colour is quite surprising. Though I knew all the species and admired them, I had no idea of the beauty of colour afforded by the varieties till I visited the Comelybank nurseries at Edinburgh a few years ago, and there found a large piece of ground covered with their exquisite tints, and looking like a most refined flower-garden. But if all this beauty did not exist, the charms of the usual form of the species, as spread out on our sunny heaths, should suffice to warrant their culture on the rockwork or among dwarf shrubs.

As for the Ericas, all are worthy of a place, beginning with the varieties of the common ling (Calluna vulgaris)—the commonest of all heaths. It has " sported " into a great number of varieties, many of which are preserved in nurseries, and these are the kinds we should cultivate. Some of them are better, brighter, and different in colour ; others differ remarkably in habit, some sitting close to the ground in dense, green, tiny bushes ; others forming fairy shrubs of a more pyramidal character, and all most interesting and pretty. These tiny shrubs and their allies in size might form a sort of edging or marginal line round a bed of choice shrubs planted in peat, as they frequently are and must be in gardens. I will merely mention the varieties pygmæa, pumila, and coccinea. Then we have the " Scotch heather " (Erica cinerea), the reddish purple showy flowers of which are very attractive, but far surpassed in beauty of colour by

a variety of the same plant called coccinea; and
there is also a white variety, as there is of Erica
Tetralix, to which is also closely related the Irish
E. Mackaiana, a plant named after Dr. Mackay of
Dublin, who found so many of the plants in Ireland
that connect its flora with that of south-western
Europe. Next we have a ciliated Heath (E.
ciliaris), a very handsome species, with flowers as
large as those of St. Daboec's heath, and the Irish
heath (E. hibernica), one of the most valuable of
all hardy plants, in consequence of its blushing into
masses of rosy red in our gardens in early spring.
It is found in some of the western counties of Ireland, and of course after it had been discovered in
other European countries. This forms a neat,
low-lying bush; grows on almost any soil, and is
one of the most valuable of dwarf shrubs; admirable for making an edging round a bed of
choice shrubs or anything else, for the rockwork or
for the mixed border. Finally, we have among
these interesting things the Cornish Heath (E.
vagans), and from what has been said of the family
it will be perceived that a very interesting bed or
group might be made from these alone. Indeed,
they would be most desirable to introduce wherever
the soil is peaty or not over arid, and might be

grown anywhere by excavating a bed and filling it with peat : but our great object should be to make the most of natural advantages, and as many persons must have gardens suited for what are called American plants, they would find it worth while to devote a spot to the British Heaths and their varieties.

Nearly allied to them we have the interesting bog Vacciniums, which may be cultivated in marshy or peaty ground. To these belong the cranberry, bilberry, and whortleberry ; and for some of these and the American kinds, people have ere now made artificial bogs in their gardens. The little creeping evergreen, Arctostaphylos Uva-ursi, or bearberry, is very neat in the garden or on rockwork. It is found in hilly districts in Scotland, northern England, and Ireland, and may be had from the nurserymen. Then the Marsh Andromeda (A. polifolia), found chiefly in central and northern England, bears very pretty pink flowers, and grows freely in a bog or peat-bed. The very small English Azalea procumbens is also an interesting native which some people try to cultivate, and where they succeed nothing can be more satisfactory, for the plant forms a cushiony bush not more than a couple of inches high. In Britain it is found only in the Scotch highlands. I have only once seen this firmly established in cultivation. Few people who admire what are called American shrubs can have failed to notice from time to time the beautiful St. Daboec's Heath (Menziesia polifolia), a plant found rather abundantly on the heathy wastes of the Asturias and in south-western France, and also in some abundance in Connemara, in Ireland. It is usually associated with American plants in our nurseries and gardens, preferring peat soil and the treatment given to such subjects. It is an elegant and beautiful plant in every way, and should be in every garden. The flowers are usually of a rich pinkish colour, but there is a pure white variety equally beautiful, while quite distinct from the commoner one. It is grown in every nursery, for its great beauty, and is therefore to be had without trouble. The very rare blue Menziesia of the Sow of Athol, in Perthshire, is also very desirable if you can get it, and I think it is sold in the Edinburgh nurseries.

The Pyrolas, or Winter-greens, are charming native plants, some of them deliciously fragrant, and all interesting, but they are difficult to cultivate. P. rotundifolia and P. uniflora are among the best, and both are rare. Should any reader attempt their culture, it will be well to bear in mind that light free leaf-mould, with sand and a little good loam, are necessary : they delight in a light spongy sort of soil, with good drainage, abundant moisture, and shade. Vinca minor and V. major are too well known to need recommendation ; there are now some finely variegated forms of the larger periwinkle, and a white-flowered kind of the smaller one is not uncommon.

One of the most precious gems in the British flora is the vernal Gentian (G. verna), which grows in Teesdale and in a few places on the western shores of Ireland. The blue of this flower is of the most vivid and brilliant description ; it is in fact the bluest of the blue, one of the most charming of all Alpine flowers, and should be in every garden of hardy plants. It may be grown well in sandy loam mixed with broken limestone or gravel, and indeed is not very particular as to soil, provided that it be mixed with sharp sand or grit, kept moist, and well drained. A very important point in the cultivation of this plant is to leave it for several years undisturbed. It is best suited for a snug spot on rockwork, where, however, it should not be placed, unless there is a good body of soil into which its roots may descend and where they will find moisture at all times. It cannot be too well known that rockworks, as generally made, are delusions—ugly, unnatural, and quite unfit for a plant to grow upon. The stones or "rocks" are piled up, with no sufficient quantity of soil or any preparation made for the plants, so that all really beautiful rock-plants refuse to grow upon them, and they are taken possession of by weeds and rubbish, which also often refuse to grow upon the "rockwork," because they cannot lay hold of it, so to speak. They are generally made either too perpendicular or too ambitiously, even in the best gardens in England—masses of rock being used merely to produce an effect, or masses of stone piled up without any of those crevices or deep chinks of soil into which rock-plants delight to root in a native state.

The right way is to have more soil than "rock," to let the latter suggest itself rather than expose its uncovered sides, and to make them very much flatter than is the rule, so that the moisture may percolate in every direction, and that the rockwork may more resemble a jutting forth of stony or rocky ground than the ridiculous half-wall-like structures which pass for rockworks in this country. I have grown this Gentiana verna very well in well-drained pots, giving it plenty of water in summer, and also in the open border in fine sandy soil, the surface being studded here and there with small stones, among and around which this lovely plant made its way and flowered " deeply, darkly, beautifully blue" every season. It is abundant in mountain pastures in central and southern Europe ; it is, in fact, a true Alpine, and may now be had in various nurseries.

The Marsh Gentian (G. Pneumonanthe) is also a lovely plant, more so perhaps than many would think this dull clime capable of producing. It should have a moist spot in a border, and is not difficult to find in the north of England ; it also grows, though less plentifully, in central or southern England. The Brighton Horticultural Society is in the habit of giving prizes for collections of wild plants, and thereby doing much harm by causing a few rude collectors, anxious to win a few shillings, to gather bunches of the rarest wild flowers, and perhaps exterminate them from their only habitats. When at one of its meetings a few years ago, I observed among the collections competing for a paltry prize large bunches of this beautiful Gentian, which had been pulled up by the roots, to form one of one hundred or more bunches of wild flowers torn up by one individual. To exhibit our wild flowers at a "flower show," where they are contrasted with hosts of Geraniums and showier subjects, is a very doubtful way of attracting people to study them; but to give prizes for the rarest plants of a locality, which in consequence are exterminated to form part of a collection of this kind, is very reprehensible. The system is bad, root and branch, and should be discouraged by every lover of wild flowers, as well as any other plan that would cause quantities of our rarest plants to be exterminated.

In the Gentian order we have also the beautiful Buckbean or Bogbean (Menyanthes trifoliata), a plant that will grow on the margin of any water or ditch or moist spot; it even grows and flowers in a moist border. It is a well-known and widely-distributed plant—everywhere over Britain, in fact; nevertheless, too much cannot be said in praise of this singularly beautiful native, with its flowers deeply and elegantly fringed on the inside with white filaments, and its unopened buds tipped with apple-blossom red. It is not often seen in a garden, though no plant, British or exotic, is more worthy of that position. It would be worthy of culture if a stove were necessary for its preservation; but, as it is accommodating enough to grow strongly under the same conditions as the water-cress, and is even less fastidious than that, there is nothing to prevent all from enjoying its beauty. Villarsia nymphæoides is also another capital water plant, with floating, small, water-lily-like leaves, which are dotted in July with a profusion of yellow flowers—so much so as to produce a very showy effect on a piece of garden water; and on such it associates very well with the white water-lily. In fact, the prettiest effect I have ever seen on any ornamental water was produced by this plant lining a small shrub-bordered bay, a group of water-lilies appearing on its outer or deep-water side. Seen from the opposite shore the effect was charming—large queenly water-lilies in front, then a wide-spreading mass of green thickly sprinkled with starry yellow, and behind all the green healthy shrubs which came to the water's edge on the shores of the little bay.

Jacob's Ladder, or Greek Valerian as it is sometimes called, also belongs to the Gentian order, and is an ornamental border plant, but its variegated variety (Polemonium cæruleum variegatum) is of the highest character and value. It resembles a variegated fern, each pinnate leaf being decidedly and well marked, and the plant forms a capital subject for edgings, or the flower garden in any way. It is much used in fine flower gardens,

and also in borders for bedding out. The
flower-stems must be prevented from rising, as it is
in the foliage that its beauty consists; and by
allowing it to flower we of course tend to prevent the
spread of the leaves and plant by the roots, or what
may in fact be called " lateral extension." Besides,
the rising flower-stems would destroy the "fern-like"
illusion. Whether British flowers are collected or
not, this will prove a decided acquisition to any
garden. Do not buy it in the form of a small and
sickly plant if you can help it, as it may "go off"
in the winter before becoming established; and
buy it or have it sent in spring—in the month of
March or April—when it may be planted in rich
light earth, and allowed to grow away at once. It
is propagated by division of the roots.

Most worthy of notice, in the Stellate or Galium
tribe, is the little white-flowering Woodruff (Asperula odorata), which bears its white flowers profusely in many British woods in spring, and I have
seen it flowering very abundantly among the trees
and shrubs round some of the Colleges at Oxford.
It should be known to every garden, in consequence
of the sweet smell it yields when dried, and kept
for a long time. There is no plant more worthy of
culture for this purpose alone, the dried stem being

as fragrant as the sweetest new hay, and continuing
to give forth its odour for a long period—an indefinite one, so far as I know.   It is fond of slight
shade, and worth planting where not found in a
wild state. When green, the "haulm" of this plant
betrays no noticeable fragrance, but begins to emit
it very soon after it is cut, and merely requires to
be placed on some dry shelf or half-open drawer,
where it may become quite dry and ready for use.

The common Red Valerian, as it is called, or
Centranthus ruber botanically, is a really ornamental garden plant, and makes a conspicuous
object on banks, borders, or large rockwork.
As it may be readily raised from seed, there can
be no difficulty in procuring it, and it should be
noted that there is a fine deep red as well as the
ordinary variety, and also a pure white, and all the
three are really ornamental plants.   Their best use
is for studding here and there on diversified or
sloping banks, in wild and half-wild spots.   They
are also useful in the picturesque garden.   Like
the Wallflower, they do well on old walls, &c., and
thus have become "naturalized" in many parts of
the country. It is the first plant that occurs wild in
newly-opened chalk-pits.

The composite or Dandelion family is generally

so ragged in appearance, that I scarcely like to introduce it here. Some unattractive members of the family are so commonly seen wherever we walk abroad, that the greatest care must be made in selecting garden subjects from it The Hieraciums are in some cases showy and fine plants. Here I will merely mention H. aurantiacum, a neat border plant, and distinct in colour, and pass on to Silybum marianum, the milk thistle ; Carduus eriophorus, a noble thistle, found chiefly in the limestone districts of the south of England—and to the great, woolly, silvery cotton-thistle, or Scotch thistle, as it is often called. These are sure to be useful, especially now, when people are beginning more to admire plants of noble or distinct form and habit. Though frequently selected as the thistle of Scotland, the Onopordum is not a native of that country ; so the Scotch thistle is a more dubious vegetable than the Irish shamrock. But, if you search the whole vegetable kingdom, you will not find among plants that are at home in our climate anything more distinct than this Cotton Thistle. A single specimen, standing in the midst or in front of green shrubs, produces a noble effect, and the plant should be in every garden. Easily raised from seed, and once established in a garden,

P

it sows itself. Then the precaution should be taken of thinning down the young seedlings, or you may have far too many of them. One isolated plant or a group or two is quite sufficient for ordinary gardens; but where there is sufficient space, it, with many other fine wild plants, might be naturalized with great advantage by simply sowing a few of the seeds in any waste or half-wild spot, or in the shrubbery. The Milk thistle, with its shining green leaves and white markings, is also very desirable among the British plants, though scarcely so much so as the great cotton or Scotch thistle.

Everywhere the common corn-flower, Centaurea Cyanus, makes a beautiful garden plant, if sown in autumn and allowed to flower with all its accumulated vigour in spring. Sown in spring, it is far inferior. I know of nothing more beautiful than a large group or small bed of the various coloured forms of Corn-flower in full bloom in spring and early summer; the bloom is so prolonged and vigorous, the flowers so pretty and so useful for the usual purposes of cut flowers. It is common, and easily had from any seedsman. One of the prettiest of all dwarf trailing silvery plants is the tomentose Diotis maritima, which is found on the southern shores of England, coming up as far as

Anglesea on the west and Suffolk on the east, but generally a rare plant in this country; it may, however, be had in nurseries, and is worthy of a place in every garden, and especially in every collection of variegated or silvery leaved plants.

The common Tansy is too coarse for any place but the herb ground, but there is a variety with leaves cut into numerous segments, and crisped up as elegantly as the New Zealand Todea superba, and this should be provided with a nook, its flowering stems requiring to be pinched off when they show. The name of this tansy is Tanacetum vulgare crispum. The double variety of Pyrethrum, now so frequent in our flower gardens, is a native plant—or, at least, the single or normal form of the species is. The Sea Wormwood (Artemisia maritima), is a neat silvery bush, freely distributed on our shores, and worthy a place in our gardens. There is a deep rose-coloured variety of the common Yarrow (Achillea Millefolium rosea)—pity one cannot avoid these hard names—which should be in every garden, and there is a very pretty double white variety of the "Sneezewort" (Achillea Ptarmica), which will be found highly ornamental. At Mr. Paul's Cheshunt nurseries I noticed it being cut extensively for wedding bouquets, during the past summer—the flowers are so purely white and neat.

Perhaps some readers may regret that I do not give the English names of all the plants, and that I do not is explained by the fact that they have no English names in a great many instances ; and would it not be a foolish barbarism to give awkward translations of the Latin names? Many people have an idea that every plant has, or should have, a " common name," whereas such only belongs to plants that have been much noticed by the people either for their beauty or "virtues." Now, as hundreds of plants are so inconspicuous, or so rare that they were never noticed till the sharp-sighted botanist took them up and gave them a Latin name, which is on the whole the best, because the language is fixed, and common to the learned of all countries, it will be readily seen why we have not English names for all our plants.   However, the next member of this natural order Compositæ, or the Daisy order, which I shall notice, is endowed with several common names, such as "Mountain Cudweed," " Cat's Ear," and " Mountain Everlasting," — the botanical one being Gnaphalium dioicum.   It is a beautiful dwarf plant, admirable for rockwork or the front of a border, or in any

way amongst alpine plants, and abounds on mountains in Scotland, Wales, and many parts of England. There is a variety called G. d. roseum, to be
had in some nurseries, that has its dwarf flowers
delicately tinted with rose ; a most desirable thing.
Neat edgings are sometimes made of this plant ;
so that there should be no difficulty in procuring
it, even supposing we cannot find it wild ; but it is
a popular plant wherever Alpines are grown, and
therefore not difficult to obtain anywhere. Gnaphalium margaritaceum is a common old plant in
gardens, its flowers having been often dried for
"everlastings," and altogether it makes a respectable, though not first-rate, border-plant,
and should be in the "garden of British wild-
flowers."

We will now turn to the extensive Harebell order,
where we shall find much beauty with little or no
raggedness—from the Harebell which swings its
bonny blue flower above the blast-beaten turf on
many an upland pasture, to the little prostrate Ivy
Campanula (C. hederacea), which is rather plentiful
in most spots in Ireland and Western England.
The giant Campanula (C. latifolia) is one of the
handsomest, and is pretty frequent. The spreading
Campanula (C. patula), of the central and southern

counties of England is also very ornamental. C. Trachelium is also good, and indeed nearly all the members of the family are of a character superior to that of most of our wild plants ; but none of them surpass in beauty the common Harebell, which, although it may look struggling for exis- tence on comparatively poor or exposed pastures and elevated spots, yet, when transferred to a gar- den, makes a vigorous plant and flowers profusely —a mass of pleasing colour. It is capital for the border or large rockwork.

The little Ivy Campanula had better be grown in a pot or peat soil, or in some moist and slightly shaded spot where it may not be overrun by tall plants. If you grow it in a pot, stand that in a saucer of water, and then the tiny Ivy-like shoots will fall down over the edge of the pot, and when dotted over with its pale blue flowers will look very interesting, especially to those acquainted with our native plants. Both this plant and the even more interesting Linnæa borealis may be grown well on the outside of the window, with a north or shady aspect during the seven warmest months of the year, by planting them in pots of peat earth. and standing these in pans of water. In winter they would be better placed in a cold frame or pit. To be able to cultivate things so interesting to the botanist, and to all who know plants, as these are, would surely be more gratifying than any amount of such subjects as we see displayed in every window.

The Ivy-leaved Cyclamen, or the common Cyclamen (Cyclamen hederæfolium), a native of Southern Europe, but not supposed to be truly British, has been found in several places, apparently wild, and as such is generally included among British plants. Being a very beautiful one, it is in all respects worthy of a place. You cannot, perhaps, easily find it wild in England, but it is not difficult to obtain, and a lovely plant it is when seen in flower. A ring of it planted round a small bed of choice shrubs forms a pretty sight, and it may be naturalized in all parts, in bare places, in woods and shrubberies. Like those of all the Cyclamens, the flowers are singularly pretty, and being densely produced in low masses, both rosy purple and pure white, they are invaluable ornaments to the autumnal garden. The Water Violet (Hottonia palustris), which bears such handsome whorls of pale purple flowers, sent up on its erect stems from its dissected leaves submerged under the water, is a choice plant for a fountain-bed or pond. Though usually supposed to grow under water, I have seen quantities of it growing most luxuriantly on soft mud-banks.

I had almost forgotten our native Primroses and Cowslips, but surely there is no need to plead for these and their numerous and beautiful varieties. The Bird's Eye Primrose of northern England—one of the sweetest of our native plants, is, however, very rarely seen in gardens. It would thrive well in wet spots on pastures and heaths, and also in bare moist spots by the side of rivulets, and in the bog bed, and on rock work, as would the smaller and beautiful Scotch Bird's Eye Primrose.

The Loose-strifes, or Lysimachias, are sufficiently ornamental for cultivation; L. Nummularia, the Creeping Jenny of the London windows, trailing its luxuriant leaves where few other plants would thrive so well. The upright-growing species L. thyrsiflora is very desirable for the margin of water, in consequence of the curious habit it has of half-hiding its flowers among the green of its leaves· A mass of it by a river, or pond, or ditch, looks very distinct and pleasing. Finally, we have in the Primula order the beautiful Trientalis of the north, a wood plant, and somewhat difficult to cultivate, but one that may be well grown in shady and

half-shady spots in peat soil—a position among
Rhododendrons etc., will do well.

Of the Thrift family, certainly the most valuable
plant is a deep and charming rose-coloured variety of
the common Thrift (Armeria vulgaris). Everybody
knows the Thrift of our sea-shores, and of the tops
of some of the Scotch mountains, with its pale pink
flowers ; but the variety I allude to is of a deep and
showy rose, and one of the sweetest things you can
employ in a spring garden as an edging plant, or in
clumps here and there in borders. This kind is
sold and known as Armeria vulgaris rubra, or A.
rubra. The common kind is not worth growing
beside it, but the white variety is. Any of the
British Statices that may be collected are worthy a
place in a collection of wild flowers. In the Goose-
foot and Dock order Atriplex portulacoides and
Polygonum Bistorta will be found the best. The
first is a silvery-looking shrubby herb, frequent on
the sea-shores ; the second a showy herb, most
plentiful in the north. Euphorbia Lathyris is the
distinct-looking and handsome Caper Spurge, which
is established here and there with us ; it is worthy
of a place, though not for the beauty of its flowers.
Nor must we forget the common Hop (Humulus
Lupulus), which I need hardly say, is very orna-

mental when well grown over a bower, or in any other position where it may have an opportunity to become fully developed.

The beautiful "Poet's Narcissus" (Narcissus poeticus), hawked about the streets of London so abundantly in spring, is generally included in native plants, though not considered truly British; but whether it be so or not, such a distinctly beautiful plant should be in every garden. The Snowflake (Leucojum æstivum) occurs in several of the south-eastern counties, and makes a handsome border bulb; the dwarf, sweet, and fine vernal Snowflake has been recently found in Dorsetshire in some abundance; while the common Snowdrop is perfectly naturalized in various parts of the country. These, it need hardly be said, should all be in any living collection of British wild flowers, and with them the Daffodil and the Wood-tulip (T. sylvestris). This last is found most frequently in some of the eastern counties of England, but may be had readily from the nurserymen, who sell it as T. florentina and cornuta. Lloydia serotina is an extremely rare little bulbous plant, found in North Wales. It is also known as Anthericum serotinum.

A Gladiolus (G. illyricus) has recently been found in the New Forest, near Lyndhurst; it is worthy of culture, and indeed is, or was, a favourite plant in many gardens before it was discovered as a British plant, having formerly been introduced to our gardens from central and southern Europe. The spring Crocus (C. vernus) is abundant in the neighbourhood of Nottingham, and other parts of England and Ireland ; and the less known but equally beautiful autumn Crocus (C. nudiflorus) is also naturalized in Derbyshire, about Nottingham, and in a few other places. It is quite needless to praise either, The blue or normal form of the vernal crocus is, or ought to be, in nearly every garden ; but the autumnal crocus is quite of rare occurrence in gardens, and should be introduced to all, because it opens its handsome flowers when most others have perished or are perishing, and closes the season of flowers so well opened by the spring crocus. It is equally easy of culture with the spring crocus, but, being so much scarcer, deserves to have a good position, good soil, and some watch-fulness, to prevent its being dug up by care-less workmen, that it may increase, and be-come a conspicuous autumnal ornament in our gardens.

The embellishment of water is really much more of an important subject than is generally supposed.

It is true that by following the directions of the
garden books, or even the best examples that we
see of water in our public gardens, nothing to boast
of can be done, but nevertheless, by a tasteful selection of really good and hardy water plants, and above
all, a judicious disposition of them, a great deal of
exquisite beauty may be produced. Hitherto this
has been very badly performed by the designers of
pieces of water, or by those who plant the margins
of them. Usually you see the same monotonous
vegetation all round the margin if the soil be rich ;
in some cases, where the bottom is of gravel, there
is little or no vegetation, but an unbroken ugly line
of washed earth between wind and water. In others
aquatic plants accumulate until they are a nuisance
and an eyesore ; and I do not simply mean the be-
low-surface weeds, like the Anacharis, but the White
Lily when it gets too profuse. Now a well developed
plant, or group of plants, of the queenly Water
Lily, floating its large leaves and noble flowers, is
an object not surpassed by any other plant in our
gardens ; but when it increases and runs over the
whole or a large part of a piece of water, and
thickens together, and the fowl cannot make their
way through in consequence, then even the queen
of British water plants becomes a nuisance. No

garden water, however, should be without a few fine
plants or groups of the Water Lily, and if the bottom
did not allow of the free development of the plant,
a lot of scrapings or rubbish might be accumulated
in the spot where it was desired to exhibit the
beauties of Nymphæa, and, thus arranged, it could
not spread too much. But it is not difficult to prevent the plant from spreading; indeed, we have
known isolated plants and groups of it remain of
almost the same size for years. Where it increases too much, reducing it to the desired limits
is of very easy accomplishment, either by cutting off
the leaves or by trimming the roots in the bottom.
The yellow Water Lily, though not so beautiful as
the preceding, is worthy of a place; and also the
little Nuphar pumila, a variety or sub-species
found in the lakes of the North of Scotland, if you
can get it. In collecting these things, the true and
the only way is to get as many as possible from
ordinary sources at first, and then exchange with
others who have collections, whether they be the
curators of botanic gardens or private gentlemen
fond of interesting plants. With a little perseverance
many good things may soon be collected in this way.

I have already (at page 206) mentioned the beautiful effect of a sheet of Villarsia nymphæoides

belting round the margin of a lake near a woody
recess, and before it, more towards the deep water,
a fine group of water lilies. The beauty of this
Villarsia is very insufficiently developed in garden
waters. It is a charming little water plant, with
Nymphæa-like leaves and numerous golden-yellow
flowers, which furnish a charming effect on fine
days when the sun is " out." It is not very commonly distributed as a native plant, though where
found generally very plentiful, and not difficult to
obtain in gardens where aquatic plants are grown.
It is in all respects one of the most serviceable of
hardy water plants.

Not rare—growing, in fact, in nearly all districts
of Britain—but exquisitely beautiful and singular
is the Buckbean or Marsh Trefoil, before alluded to,
with its flowers elegantly and singularly fringed on
the inside with white filaments, and the round unopened buds, polished on the top with a rosy red
like that of an apple blossom. In early summer
when seen trailing on the soft ground near the
margin of a stream, this plant is very beautiful, and
should be grown in abundance in every piece of
ornamental water. It will grow in a bog or any moist
place, or by the margin of any water. Though a
rather frequent native plant, it is not half sufficiently

grown in garden waters ; but, indeed, these are invariably neglected.

If you have ever seen the Flowering Rush (Butomus umbellatus) in flower, you are not likely to omit it from a collection of water plants, as it is conspicuous and distinct. It is a native of the greater part of Europe and Russian Asia, and dispersed over the central and southern parts of England and Ireland. Plant it not far from the margin ; it likes rich muddy soil. The common Sagittaria, prevalent, very prevalent in England and Ireland, but not in Scotland, might be associated with this; but there is a very much finer double kind to be had here and there, and which is probably a variety of the common kind. It is really a fine plant, its flowers being white, and resembling, but larger than, those of the old white double rocket. It grows in abundance in the tea or pleasure gardens of the Rye House at Broxbourne, where it fills a sort of oblong basin or wide ditch, and looks quite attractive when in flower. Its large tubers, or rather receptacles of farina, are frequently discovered and destroyed by wild fowl, which suggests that it might be worth planting as food for such birds.

Among bold and picturesque plants for the

water-side, nothing equals the great Water-dock (Rumex Hydrolapathum), which is rather generally dispersed over the British isles, and has leaves quite sub-tropical in aspect and size, becoming of a lurid red in the autumn. It forms a grand mass of foliage on rich muddy banks. The Typhas must not be omitted, but they should not be allowed to run everywhere. The narrow-leaved one is more graceful than the common one. Carex pendula is very fine for the margins of water, its elegant drooping spikes being quite distinct in their way. It is rather common in England, more so than Carex Pseudo-cyperus, which grows well in a foot or two of water, or on the margin of a muddy pond. Carex paniculata forms a strong and thick stem, sometimes three or four feet high, somewhat like a tree-fern, and with luxuriant masses of falling leaves, and on that account is transferred to moist places in gardens and cultivated by some persons, though generally the larger specimens are difficult to remove and soon perish. Scirpus lacustris (the "Bulrush") is too distinct a plant to be omitted, as its stems, sometimes attaining a height of more than seven and even eight feet, look very imposing; and Cyperus longus is also a desirable thing, reminding one of the aspect of the Papyrus when in flower. It is found in some of the southern counties of England. Cladium Mariscus is also another distinct and rather scarce British aquatic which is worth a place.

As for the Ferns, it is needless to mention them, considering the immense attention that has been paid them of late years. Whole nurseries are now almost exclusively devoted to the production of British ferns and their varieties. My object is to encourage the culture of things that are comparatively neglected, and however graceful and beautiful ferns may be, and however indispensable the fernery, as an adjunct to the flower-garden, my readers have but to attempt the culture of the handsome British flowering-plants, combined, if the cultivator so desires it, with the best alpines, spring flowers, and herbaceous plants of all countries, to find infinitely more enjoyment therefrom than ferns are capable of affording.

But though ferns are not in need of advocacy, their allies the Equisetums are, some of them being of graceful and distinct habit. One of the most strikingly distinct and elegant plants in the Oxford Botanic Garden grows profusely along by the wall, in the shady fern border, in that very old and most interesting botanic garden. It is the British Equi-

Q

setum Telmateia, or "Great Equisetum," which grows pretty commonly in the greater part of England and Ireland, attaining its greatest development in rich soil and in shady spots. It there attains a height of three or four feet, and the numbers of slender branches depending from each whorl look most graceful. It should be planted in a shady place, near water if convenient, but it thrives famously in deep moist soil, in any position in a garden where ferns thrive, and as it associates well with them, in or near the fernery will be found a good position for it. The wood Equisetum (E. sylvaticum) common all over Britain, is considerably smaller than the preceding, but even more graceful ; indeed, sufficiently so to warrant its being grown in pots, though it thrives well in any shady moist position. The long simple-stemmed Equisetums, or Horse-tails, are also interesting to cultivate in wet or marshy spots, or by the sides of water, but are not so graceful or ornamental as the species above-named, which are as well worth growing in a garden as the costliest productions of tropical climes, which entail endless work and a perpetual cost to maintain.

Passing by the numerous British Willows and the few British Pines, we come to the interesting

order of Orchids—everywhere beautiful and singular, whether gorgeously developed, as in the hot
or moist East, or small and tiny on the Kent and
Surrey hills, where the Bee Orchis produces its
peculiar flowers so abundantly. Now, many of these
small British Orchids are, in my opinion, as pretty
as many of those cultivated in the stove or greenhouse. It is most interesting to see and to collect
them when wild, and still more so to cultivate
them. If you can succeed in growing the British
Orchids, you are not likely to fail with any other
hardy plants. They are the most difficult of all to
cultivate, but amongst the most interesting things
which can be grown. In many or most parts of the
country the Bee Orchis and some other rare ones do
not grow ; how interesting it would be in such districts to have the Bee Orchis to show in one's garden!
I have never seen our Orchids grown in more than
half a dozen gardens, but, nevertheless, have no
doubt that they can be very well grown therein, because I have cultivated the Bee Orchis and the Fly
Orchis and the Hand Orchis, and a number of other
British Orchids, for several years, and flowered them
annually too. People generally make a mistake
by putting them in pots. If the plants should
make a good attempt to grow there, the long fleshy

roots that some species produce have no chance of finding a suitable, steady medium in which to thrive. The pot with its soil is liable to vicissitudes from want of water, and from the hot dry air of summer always playing upon its porous surface. Therefore, though pots are the usual resource even in a botanic garden, Orchids never do well in them, but usually live for a year or two and then perish.

I succeeded with them by devoting a small bed to their culture, in a somewhat open but sheltered spot. The first thing I did was to dig some chalk into the bed, so as to give the plants the constituent in which they are found most abundantly. Of course, I should not have had to do this if the soil were chalky; and as numbers of my readers must have gardens upon chalky soils, I may assure them that they will have no difficulty in growing the choicest British Orchids. Then I planted the various kinds, and succeeded with every one of them except the parasitic one, which, indeed, it was vain to attempt. I allude to those kinds that are parasitic on the roots of trees, though apparently depending on their own roots in the ground.

My only difficulty was to imitate, to some extent, the state of the surface of the ground which exists where they live in a wild state. I knew that the surface dressing of stunted, storm-beaten grass among which they nestle prevents the ground from cracking, hinders a good deal of evaporation, and also shelters the plants in winter—in short, keeps the surface open, natural, and healthy. To plant grass over a bed in a garden would never do, because the shelter and richness of the ground would induce it to grow so strong that unless we were to look after and shorten it very frequently there would be no chance of keeping it within bounds; and if we did not do that, it would soon smother all the Orchids. I found a substitute in cocoa-fibre mixed with a good sprinkling of silver sand and a little peat to give it some weight and consistency. An inch or two of this material was spread over the bed, and it succeeded perfectly in answering my ends, *i.e.*, it prevented cracking and evaporation, and kept the surface in an open, healthy state. Of course I inserted the plants firmly and without injuring their roots—a great point. Few people know how to plant anything beyond a strong bedding plant.

If one of these Orchids which are accustomed to send their fleshy roots down below moist accumulations of broken chalk in search of food, were to be planted like a bedding plant, it would soon perish. I made the ground quite firm, then cut a straight deep little trench with a straight trowel, and against the flat side of this little cut placed, lengthwise of course, the spreading hand-like roots of my Orchids, pressing the soil firmly but gently against them, and being particular that the "neck" or collar of the plant was nicely pressed round and firm—a thing that is worth attending to in every case of planting. If you examine a plant after some people have inserted it, you will find the whole of the top of the ball loose, and perhaps uncovered by soil—a state most conducive to an early death or stunted growth if the weather prove dry; therefore always plant firmly, and try and place the roots and neck of the plant as much as possible in the condition that plants enjoy in a wild state.

Well, in this way I have grown and freely flowered the most curious and beautiful Bee Orchis, the Spider Orchis, the Fly Orchis and a dozen others less difficult to cultivate. The marsh Epipactus palustris is one of the easiest native Orchids to cultivate, growing well in an artificial bog or moist border; whilst most of the Orchises will do well under the treatment above described. The Bee, Fly, and Spider Orchids belong to the genus Ophrys.

The common Orchis maculata, found almost everywhere in the British islands, is one of the freest to
grow in a garden ; it makes large tufts of the
greatest beauty in a stiff good loam, and I have
found it grow with ease in almost any position.
It wants no chalk, though it does not refuse to
grow in it.   The best wild spread of it I have
ever seen was in some meadows in Buckingham-
shire, where there was a strong bloom of this
sweetly-coloured Orchid for almost every flower-
head of grass in the fields ; and I need hardly say
the effect was of the most beautiful description.

Lately nurserymen have been offering a plant
which they think a variety of this, under the
name of O. maculata superba.   This is in reality
the true British Orchis latifolia, a noble species,
easy to grow in a moist spot, and producing long
spikes of bloom.   O. militaris and O. fusca are
among the handsomest species ; but all are interesting, even when not pretty, from the early
spotted O. mascula to the Butterfly Orchis, both
of which are of easy culture in a garden.

Perhaps the rarest and finest of all the British
Orchids is the Lady's-slipper, nearly extinct, but
still probably to be found in the North, though too
rare to be looked for in the hope of transferring it

to the garden. Some of our nurserymen supply it, and they get their supplies from the Continent, where it is a widely distributed plant. It should be planted in broken limestone and fibrous loam, on the eastern side of a rockwork. When well grown it is a beautiful plant, quite as much so as some of the Cypripediums grown in the Orchid house, but, being perfectly hardy, is of course far more interesting and suitable for the British garden. The most important thing with regard to the Orchids is the procuring of them in a suitable state for planting. When they are gathered in a wild state, the roots should be taken up as carefully as possible, and transferred to their garden home quickly and safely. They are very often sold cheaply in Covent Garden, but the roots are generally mutilated, not only from careless and shallow taking up, but from being so tightly bound round with moss and matting that any bit of root they had when taken up is bruised to death. I got a capital stock by finding one of the men who collected ferns and wild plants for Covent Garden, showing him the kinds I wanted, and telling him not to bind them up individually, but to lay them in loose layers in moss, having taken them up carefully, with the roots or tubers entire. Of course, if

I lived near localities in which the rarer and more interesting Orchids are found, such as many parts of Kent and Surrey, I should gather them myself, using a very strong spade, or instrument, to get them well up out of the firmly-bound chalky earth.

Among native bulbs there are some very interesting. The Snake's Head (Fritillaria Meleagris) is abundant in some parts of the south and east of England, and it is in all respects worthy of the best attention in a garden, though it requires very little beyond being planted and allowed to grow away undisturbed. I know of nothing prettier in the Spring Garden than the singular suspended bells of the English Fritillary, often so prettily spotted, and occasionally white. The white form is sometimes called F. præcox, and being of a good white, it is a most desirable plant to encourage in every garden, the large, white, drooping bells being so distinct from most other hardy plants that flower at the same season. Of the British Alliums, A. triquetrum, found somewhat abundantly in the island of Jersey, is best worthy of a place, its white flowers striped with delicate green being pretty. The two British Squills, though not so ornamental as some of the Continental species, so conspicuous among spring flowers, must not be forgotten in a full collection, nor the varieties of the wood hyacinth, and there are several of interest, both white and pink. The Two-leaved Lily-of-the-valley (Convallaria bifolia) is a diminutive and sweet little herb, found in only a few localities. In Lord Mansfield's woods, near Hampstead, I gathered it a few months ago, and it is abundant there in a well-shaded spot. It does well either on border, or rockwork, or in the wilderness. It is common on the Continent, and may be readily had from some nurseries, and in all botanic gardens in this country.

The common Lily-of-the-valley is a true native plant, abundant in some counties, though wanting in others. It is surely needless to recommend it to my readers as a garden ornament, but I may suggest that it might be "naturalized" in many woods and shrubberies with the best effect—it is so interesting to meet with things like this in an apparently wild state. The handsome, graceful Solomon's-seal (Polygonatum multiflorum) and the Lily-of-the-valley should be planted to establish themselves in a wild or semi-wild state in every place which possesses the smallest resemblance to a shrubbery or wood ; nothing can be more grace-

fully beautiful than the flower-laden stems of the
Solomon's-seal arching forth from such positions. It is not enough to meet with the Lily-
of-the-valley in the garden, we should meet with it
in the wilderness, by the woodland walk, among
the Primroses and Bluebells, and wherever native
or hardy plants are cultivated. The Star of
Bethlehem (Ornithogalum umbellatum) and the
drooping Ornithogalum are established in several
parts of the country. The first is a well-known old
garden plant; the second a handsome kind with
drooping flowers. With these I will group the
Meadow Saffron (Colchicum autumnale), which is
abundant in some parts of Ireland and England,
and frequently cultivated as a garden plant, commonly under the name of the Autumn Crocus, which
name properly belongs to the blue Crocus nudiflorus before alluded to. Both plants should be
associated in every garden.

In the rather large Grass family the common
Ribbon Grass (Phalaris arundinacea variegata),
Hierochloe borealis (a rare northern plant, sweet-
scented when dried), Milium effusum (a handsome
wood grass, rather common), the exquisitely graceful Apera Spica-venti of the eastern counties, the
Hare's-tail Grass of the Channel Islands (Lagurus

ovatus), the Quaking Grasses (Briza media and minor), the well and strikingly-variegated Cocks-foot grass (now beginning to be extensively used as a bedding plant), the new variegated variety of the common Poa trivialis, and Elymus arenarius, a strong conspicuous, glaucous grass, are worth growing, being for the most part either very interesting or beautiful. Some of those Grasses, now never seen in a garden, are worthy of being grown for dinner-table decorations, to which they would add as much grace as any costly exotics.

We have thus glanced at the garden of British Wild Flowers, from showy, self-asserting Butter-cups and Poppies to modest Grasses of exquisite grace ; but it would require much greater space to do justice to the numerous and delightful aspects of vegetation they are capable of producing : to enjoy these the best way is to set to work and form them.

THE END.

Printed in the United States
By Bookmasters